D0217974

Table of Contents	Pages

To the Student

Welcome! I hope you find this Student Activity Manual helpful as you begin your tour through the world of chemistry. Of course, this tour is not like any you may have taken before of a country or city. Instead, this trip takes you deep inside the objects that make up our world so—that you can understand how they work in terms of the submicroscopic particles that make them up. With that in mind, you can think of me and this manual as tour guides on your trip. Like any experienced tour guide, I love taking this trip (I hope that comes across to you), but I also remember what was confusing about it and what I struggled to understand my first time. So our tour will start off at the very beginning and will make sure you have a strong foundation before we move on.

Now, if you wanted to truly understand a new country, it might be helpful to read some books about it. However, the best way to learn is to actually explore its cities and towns and speak to the people who live there. It's the same thing with the world of chemistry. The best way to learn is to dive in and get your hands "dirty." You need to make your own observations and struggle with your own interpretation of those observations. The work can be hard, but the rewards are great. By the end of our trip, nothing around you will look the same! You'll contemplate solutions when you drink a cup of coffee, you'll ponder electrochemistry when you use a battery, you'll think about exothermic reactions when you put gasoline in your car, and you'll consider kinetics when you take your milk out of the refrigerator.

Of course, if you want to get the most out of this tour, you need to do some preparation. Make sure you do the assigned textbook readings and Pre-Activity Problem Sets before each class and spend some serious time working on your Post-Activity Problem Sets. The problem sets have been written to make sure you are getting the important points from your trip. And while you are on your trip, try to have fun and stretch your imagination. Each experience has been designed to show you another amazing aspect of the world of chemistry. Thousands of students have successfully taken this tour with me and if they can do it, so can you. You just have to be observant and want to better understand the world around you. With any luck, this first trip will be a starting place for further tours on your own or in other science courses.

Happy explorations,
Jeff Paradis
California, January, 2010

Acknowledgments

I am grateful for the invaluable help provided by my colleague Patrick Sparks in developing early drafts of several of the activities in this manual. The puzzle pieces in "Molecules and Lewis Dot Structures" are his invention.

Several research students contributed to this activity manual. I'd like to thank Holly Garrison for her work on early drafts of "Naming Ionic Compounds." It was her idea to use games to make naming ionic compounds fun. I'd also like to thank Chad Eller for developing and testing "Kinetics and the Rates of Reactions."

I'd like to acknowledge the work of Ronald J. Gillespie, James N. Spencer, and Richard S. Moog. The article "Demystifying Introductory Chemistry Part 1: Electron Configurations from Experiment" (*The Journal of Chemical Education*, Volume 73 Issue 7, July 1996: pp 617 – 622) was the basis for using ionization energies to develop the shell model in "The Electrons and the Shell Model."

This manual would not have been possible without the financial support of the College of Natural Sciences and Mathematics at California State University, Sacramento. Dean Marion O'Leary's enthusiastic support for faculty members interested in exploring nontraditional methods of instruction was invaluable to this project.

I'd also like to thank my parents who always stressed the importance of a good education, Satoshi Okuno for tirelessly proofreading the activities (several times) and for his never-ending support, and the Salty Pretzels for making work so much fun.

Finally, I would like to express my appreciation to the scores of students who, every semester, have given their feedback so that the activities in this manual could improve. I can only hope they all learned as much from me as I learned from them.

—Jeff Paradis
California, January 2010

This material is based upon work supported by the
National Science Foundation under Grant number (DUE-9953181).

Explorations in Conceptual Chemistry: Introduction

To the Instructor

This activity manual is designed to be flexible and to meet your needs. However, the following nine activities are considered to be "core concept" activities and are crucial to your student's conceptual development. Additional activities should be selected depending on your students' needs and interests.

Content Overview

The selection of course content was determined by considering three requirements: the need to include fundamental chemistry topics, the desire to address the scientific literature concerning student naïve conceptions, and the goal of having students build a conceptual framework to serve as a foundation for their future learning. The content in thIS manual covers many of the topics traditionally dealt with in an introductory chemistry course (atomic structure, molecular and ionic bonding, acids and bases, solutions, and balancing chemical reactions), but it does so in a novel way. The order in which the concepts are presented was chosen so that the connections between the concepts is clear and leads to a strong conceptual framework. The end result is that this activity manual is centered on several "big ideas" concerning various aspects of the particulate nature of matter. While developing their model of matter, students explore the structure, properties, and interactions of the submicroscopic particles that make up our world. The emphasis will be to understand a few topics in depth rather than have a shallow overview of many topics. Research in cognitive theory suggests that skills developed while becoming an "expert" in a given topic are transferable to new learning situations. Whenever possible, conclusions concerning the submicroscopic world are based on direct student observations of macroscopic phenomena. Over the course of the semester, students refine their model of matter. Many of the topics, such as the treatment of the states of matter, are revisited at increasingly more advanced levels using a cycle approach.

Even when it is not possible to have the students use direct, personal observations to gather data (such as with the development of a model of the internal structure of the atom), students are provided with information (such as photoelectron spectroscopy results) that still allow them to draw the conclusions firsthand. All of the activities are chosen to be safe, use readily available materials, and clearly demonstrate the key concepts. Each activity in this manual contains a Post-Activity Problem Set, written to require students to apply their knowledge to new scenarios and to integrate it with previous knowledge.

Pedagogy and Suggested Use

To make the best use of the laboratory time, pedagogical choices were made, with the goal of providing students with the most effective learning environment for their own studies, while explicitly modeling best teaching practices in science education. A constructivist approach is the cornerstone for this manual's "learn-by-doing" model. At every level throughout the course (whether it is hands-on exploration of the nature of solids, liquids, and gases or developing the shell model of the atom, using ionization energy data), students are directly engaged with the observations and data being used to develop their conceptual framework.

Though the general approach employed in this activity manual is that of guided inquiry, rather than forcing all the activities into one format (for example, all discovery based or all expository), this activity manual was written to let the content determine the most effective pedagogical approach for each activity. Regardless of whether the pedagogy behind the activity is classified as guided inquiry, constrained inquiry, collaborative learning, peer teaching, or model-based learning, the key is that the activity is always student centered.

The easiest way to illustrate the incorporation of best practices is to run through a typical activity cycle from start to finish. Each activity begins with a student cover page with (1) a list of key concepts related to the activity, (2) specific learning objectives students should be able to meet after the completing the activity, and (3) directions for the student to complete before coming to the activity period. Included in these requirements is an assigned textbook reading and a page of Pre-Activity Problems. These problems tend to require low-level cognitive input and are designed primarily to ensure that the student has completed the required reading. Requiring students to take ownership of their learning by coming to lab prepared for the activity is a fundamental aspect of the course philosophy, which places the learning process squarely on the shoulders of the student.

Having students come to class prepared also serves to cut down almost entirely on the need for any pre-lab lecture. The Pre-Activity Problem Set is checked off for a small number of effort points (usually five points) at the start of lab (late students loose these points) and then is immediately passed back to the students. This should only take 2 – 3 minutes to do. I have students turn the Pre-Activity Problem Sets into a box at the start of class and I typically have them all recorded and passed back before class even starts.

The class then spends between 5 – 10 minutes having a student-centered discussion about any confusing points from their reading or from the Pre-Activity Problem Sets (instructor answer keys are available through your Pearson/Prentice Hall book representative). Experienced instructors are often able to use this discussion time to begin addressing any student misconceptions associated with the underlying concepts in that activity. Pre-activity lectures, if given at all, tend to be brief and may deal with "big-picture" contexts, pointing out underlying pedagogy or discussing any possible safety issues (open flames, dry ice, etc.) related to that activity. If I feel students are not coming to class prepared and cannot carry the discussion, I reserve the right to give them a pop quiz where they have to earn the Pre-Activity Problem Set points instead of just getting them for effort. I never have to do this more than once a semester before they start coming prepared.

Working in pairs, the students then carry out the activity for that day. Many of the activities are set up as self-paced stations where students are led with guiding questions to carry out an activity that not only engages them, but encourages "doing with understanding." When the students are done, I again lead a student-centered discussion through each of the Parts (again, answer keys are available to instructors). The wrap-up is crucial for making sure that the students all arrive at the same understanding of the key concepts covered in that activity. Some activities are more difficult and I have found that it is best to have a brief discussion after students complete each part so that we can make sure everyone is on the same track before they continue too far. At our school, we have 1-hour-and-40-minute lab periods. Most of the activities take between one and two lab periods.

At this point, the students are now ready to apply their understanding to new scenarios in the Post-Activity Problem Set, which is given as homework and is due the following period (instructor answer keys are available). Unlike the Pre-Activity Problem Sets and the questions that go along with the various parts of the Activity, the Post-Activity Problem Set is actually collected and fully graded. I don't collect/grade the other things in order to cut down on grading and because the student-centered discussions give everyone (me and the students) immediate feedback. I can see if the class is ready to do the Post-Activity Problem Set on their own or if I need to clarify something. The Post-Activity Problem Sets are designed to evaluate student mastery of the testable learning objectives and to provide students with feedback on questions similar to those they will see on exams and quizzes.

Unit 1: The Particulate Nature of Matter

Everything around you is made up of submicroscopic particles. That means your clothes, the air you are breathing, the walls of the room you are in, this activity manual, and, yes, even *you* are all made up of particles that are too small to directly see. With that in mind, it is no wonder that chemistry can seem intimidating... here you are getting ready to study things that are too small to even see! However, if you approach it with the right attitude, that is where the fun actually lies. Since the particles that make up matter are submicroscopic, chemists are required to play detective. Chemists are constantly making observations and using those observations to draw conclusions about things that can't be seen. But you don't need to take my word for it. In Unit 1, you will begin to fine-tune your powers of observation by looking around you and realizing that you can explain some simple everyday occurrences in terms of submicroscopic particles. The skills and knowledge you gain in Unit 1 will serve as a foundation for all later learning in chemistry.

Activity 1a: Matter Is Particulate (Low-Tech Evidence)
COVER SHEET

Before Class
- Read the pages in your textbook dealing with the "definition of the term *matter*" and the "origin of the word *atom*." Your instructor may assign specific pages for you to read.
- Complete the Pre-Activity Problem Set on the next page.

Key Concepts
- Matter is composed of particles that are too small to see, even with a microscope (i.e., they are submicroscopic).
- Many day-to-day observations can be explained if we consider that matter is composed of submicroscopic particles.

Learning Objectives
- You will be able to define the term *matter* and classify various things as to whether they are or are not examples of matter.
- You will be able to summarize the early history of the term *atom*.
- You will be able to explain a variety of everyday phenomena (i.e., macroscopic observations) in terms of the particulate nature of matter.
- You will be able to utilize both written and pictorial representations to communicate your understanding of the particulate nature of matter.

Name: _____   Score: _____ points

Answer the following questions based on your assigned textbook reading.

1) What is the scientific definition of the word *matter*?

2) List three things that are examples of matter.

3) List three things that are NOT examples of matter.

4) Who coined the term *atom*? Roughly when was it first used? What does the term mean in Greek?

5) What was a major competing theory in ancient Greece for the atomic theory of matter?

6) In this class we will use the phrase *submicroscopic particle* for several activities before formally introducing the term *atom*. In order to not misuse the term *atom*, you should avoid using it until then. Why do you think we start off using *submicroscopic particle*? What is it about the term that is more self-explanatory (or descriptive) than the term *atom*?

Instructions: Soak a <u>cotton ball</u> with <u>alcohol</u> so that it is thoroughly wet but not dripping. Drag the cotton ball across your <u>desk or the blackboard</u> to make a wet streak about 1 foot long. Use the space below to record your observations over the next 5 – 10 minutes. Dispose of the used cotton ball according to your teacher's directions.

Observations:

Questions:
1) Without using technical terms, write a short statement describing what is occurring on the submicroscopic (particulate) level as the alcohol "disappears." Your statement should be consistent with both your macroscopic observations and with the idea that matter is made up of particles that are too small to see. Your statement should be based only on what you can directly conclude from your observations, not on previous knowledge.

2) How does "The Case of the Disappearing Alcohol" support the theory that matter is made up of submicroscopic particles? As part of your answer, explain how the observations you made during this activity would look different if the alcohol was simply one continuous macroscopic object that was not made up of submicroscopic particles.

Instructions: Fill a <u>graduated cylinder or tall glass</u> with <u>water</u>. Let it sit for a minute so any bubbles may rise and the water is clear. Prop a piece of <u>white paper</u> up behind the graduated cylinder so it will be easier to make your observations. Using a <u>spoon or spatula</u>, add some <u>darkly colored powdered drink mix</u> (about the size of a pencil eraser) to the center of the water in the graduated cylinder. Do NOT stir. Use the space below to record your observations until no more solid is visible (about 5 – 10 minutes). Do NOT drink the colored water. When you are done, dispose of the colored water according to your teacher's directions.

Observations: Draw and label a series of three pictures that represent your macroscopic observations at the beginning, middle, and end of the "Mystery Crystals" activity.

Beginning	Middle	End

Questions:

1) Have the submicroscopic particles that made up the solid drink mix disappeared when you added them to water? Explain your answer.

2) Describe what happens on a submicroscopic level to the solid crystals of drink mix as they are added to and fall through the water. Support your answer with as much macroscopic evidence as possible.

Part C: Magic Balloon

Instructions: A few drops of a <u>liquid</u> with a familiar odor have been carefully put inside a <u>balloon</u>. The balloon was then blown up and the opening was tied off. Pick up the balloon and smell it. Record your observations in the space below.

Observations:

Questions:
1) Draw and label a picture that uses submicroscopic particles to explain your observations.

2) Imagine you repeat this experiment with a Mylar balloon (the shiny, metallic-looking ones that are filled with helium and tend to last a long time). Predict what you think will happen and explain the basis for your prediction.

Instructions: Pour approximately <u>20.0 mL of water (colored with blue food coloring)</u> into one <u>graduated cylinder</u> and approximately <u>20.0 mL of alcohol (colored with yellow food coloring)</u> into a second <u>graduated cylinder</u>. Record the volumes of each liquid below. Record your prediction for the total volume after adding the alcohol to the water. Pour the alcohol into the graduated cylinder with the water and stir carefully with a <u>stirring rod</u>. Record the volume and color of the water/alcohol mixture below. When you are done, dispose of the water/alcohol mixture according to your teacher's directions.

Prediction/Observations:

Volume of Water (Blue):   Volume of Alcohol (Yellow):

Predicted Volume of "Water + Alcohol:"   Actual Volume of "Water + Alcohol:"

Color of the "Water + Alcohol:"

Questions:

1) What was unexpected about your observations?

2) Following the example below, come up with two more possible explanations for your unexpected observation. How could you go about testing each possible explanation? For each test, what would you expect to happen if the explanation was correct? If the explanation was incorrect?

Possible Explanation	Test	Result if Explanation Is Correct	Result if Explanation Is Incorrect
Example: "Bob" was reading the volumes incorrectly.	Have "Susie" try reading the volume.	"Susie" might get a different volume reading than "Bob."	"Susie" might get the same volume reading as "Bob."
1)			
2)			

Instructions: Your instructor will show you a demonstration that will help you understand the "Mysterious Mixing Liquids." In the space below, summarize what your instructor did.

Observations:

Questions:
1) Based on your observations, explain why the actual volume of the metal balls did not match the expected volume of the metal balls.

2) Based on this demonstration, go back and explain the unexpected results you observed with in *Part D* with the "Mysterious Mixing Liquids."

3) One limitation of this analogy is that there are submicroscopic particles (i.e., air particles) between the metal balls in the demonstration but not between the water particles. What is between the submicroscopic particles of water? Draw two pictures illustrating this difference. Since you will be drawing pictures with a mixture of macroscopic (metal balls) and submicroscopic particles, be sure to label which is which.

Labeled illustration of macroscopic metal balls and the submicroscopic air particles between them.	Labeled illustration of the submicroscopic particles of water and the _____ between them.
beaker of metal balls	beaker of water

Students are often concerned about their first graded work. "What is the teacher looking for?" and "What are his/her expectations?" are common questions students have. In order to help alleviate anxiety about the assessment process, here is some practice for you. Below you will find three sample questions in **bold** along with actual student responses. The student answers are considered high quality and are of the type you should strive for. If they can do it, so can you! It just takes practice. Note that good answers are NOT necessarily long. They tend to be clear and concise, answer all parts of the question, and back up statements with evidence. These student answers are descriptive and avoided using technical terminology. Consider covering the answer and trying to write your own. Then compare it to the student response.

Sample Question #1: **Scientists often draw conclusions concerning the submicroscopic world based on their observations of the macroscopic world. What is the difference between an observation and a conclusion? Give an example of a conclusion you arrived at in this activity and the specific observations you used to defend that conclusion.**

Student Answer to Sample Question #1: An observation is the recorded results of what was seen, heard, tasted, measured.... The conclusion is how one tries to explain or make sense of the observations. For example, we observed that when a streak of alcohol is made on a desk, it slowly gets smaller while leaving the smell of alcohol in the air. Based on this, we concluded that the alcohol is made up of submicroscopic particles that are too small to see and that have gone from being together in a puddle on the desk to separating and filling the air.

Sample Question #2: **A cat walks through your yard. An hour later a dog walks through your yard with his nose to the ground, following the path of the cat. Explain how this simple observation supports the model that matter is composed of submicroscopic particles. If the particles making up matter were macroscopic, how might the path followed by the dog be different?**

Student Answer to Sample Question #2: The observation suggests that the dog is following a trail left behind by the cat. The particles making up the trail are being left behind as the cat rubs against the grass and dirt. These particles are too small to see (i.e., submicroscopic), but can be detected by the dog's keen sense of smell and are used to identify the cat and the path it took. If there were no submicroscopic particles, the cat either would not leave any particles (no trail) or the particles would be visible (like leaving a trail of bread crumbs).

Sample Question #3: **A small bottle of perfume is opened and the scent quickly fills the air. By the next day, however, the bottle is empty and the perfume can no longer be smelled. Have the molecules in the perfume disappeared? Explain why the bottle is empty and why the perfume cannot be detected.**

Student Answer to Sample Question #3: No, the perfume has not disappeared. Perfume is an example of matter so it is made up of submicroscopic particles. We can smell perfume because some of the particles enter our nose. Over time, the bottle becomes empty as all of the particles leave the perfume and move into the air and spread out into the room. The perfume can no longer be detected because the individual particles in the air are too small to see and are too spread out to smell.

Name: _____   Score: _____ points

These questions were designed to be challenging applications of what you learned in this activity. They will provide you with practice answering the types of questions you will see on exams. This is where you prove to your teacher and to yourself that you know the material, so take your time and give each question your best effort!

1) Propose a submicroscopic-level explanation for what is happening when you use a tea bag to make a cup of tea. Begin by defending your explanation with as many macroscopic observations as possible. Imagine putting all of your senses to work!

    Macroscopic Observations:

    Submicroscopic Level Explanation:

2) In *Part D* of this activity, we saw blue and yellow submicroscopic particles seemingly turn into green particles. Are the blue and yellow particles still there or have they really turned into green particles? If they are still there, then why do they appear green? Explain your answer. Hint: Take a close look at the screen of your computer or cell phone. What do you notice about how the small, individual pixels appear when they are viewed from a distance?

Continued on Back →

3) You have balloons made by two different companies. They look the same, but you decide to try some tests with them. Imagine that you fill one balloon from each company to the same size; the only difference is that one balloon is filled with helium and the other is filled with air. The next day, the helium balloon is smaller than the balloon filled with air. In the spaces below, propose two explanations for this observation: one that is related to possible differences in the material making up the balloons and one that is related to possible differences in the contents (helium versus air) of the balloons. Next, come up with a possible test that could help you determine which of your two explanations is most likely. Clearly describe your proposed test and the expected outcomes if the explanation was due to the differences in the balloons versus the differences in the contents.

Explanation #1 (Due to Differences in The Material Making Up the Balloons):

Explanation #2 (Due to Differences in the Contents of Balloons):

Possible Test to See Which Explanation Is Correct:

Expected Outcome if Explanation #1 Is Correct:

Expected Outcome if Explanation #2 Is Correct:

Activity 1b: Matter Is Particulate (High-Tech Evidence)
COVER SHEET

Before Class

- Read the pages in your textbook dealing with "scanning tunneling microscopy (STM) or atomic force microscopy (AFM)" and "physical and conceptual models." Your instructor may assign specific pages for you to read.
- Read pages titled "How an Atomic Force Microscope (AFM) Works" and "Photographs of the Setup of an Atomic Force Microscope (AFM)" on pages 19 – 20 of this manual.
- Complete the Pre-Activity Problem Set on the next page.

Key Concepts

- Scientists use models to represent complex ideas or mechanisms (conceptual models) and to show objects on a more convenient scale (physical models).
- The particulate nature of matter is consistent with the results obtained from modern experimental techniques.

Learning Objectives

- You will be able to explain, in simple terms, how AFM works.
- You will be able to differentiate between physical and conceptual models.
- You will be able to explain why models are used in chemistry.
- You will be able to use various models to demonstrate how AFM works.

Name: _____   Score: _____ points

Answer the following questions based on your assigned readings.

1) Why do we need to use models in chemistry?

2) Why can't we build a really powerful microscope that will allow us to directly see the submicroscopic particles that make up matter?

3) What causes the AFM probe to move up and down as it scans the surface of the material being studied?

4) What is the role of the laser in an AFM experiment?

5) Why is the AFM instrument sometimes placed in a plexiglass box and hung from a tripod during an experiment?

How an Atomic Force Microscope (AFM) Works

The theory that matter is made up of submicroscopic particles is fundamental to our current understanding of chemistry (as well as aspects of biology, physics, astronomy, engineering, and geology). Until recently, however, scientists had only indirect evidence for the existence of these particles (see *Activity 1a*). Being able to see these particles would allow scientists to, among other things, design better computer chips and drugs. You might wonder why we can't just make a microscope powerful enough to look at these particles. It turns out that the wavelength of visible light is too large to resolve the distances between individual particles. This is analogous to trying to read Braille while wearing a baseball glove.

In the early 1980s, scientists at IBM developed a method for making images of the surface particles of certain materials. They named this method scanning tunneling microscopy (STM) and were awarded the 1986 Nobel prize for their groundbreaking work. Never before had scientists been able to enter the submicroscopic world with such ease and clarity. However, STM can only be used to study materials that conduct electricity.

To overcome this limitation, atomic force microscopy (AFM) was developed. With AFM, a sharp probe is attached to the end of a flexible cantilever (see figure below). The probe and cantilever are then scanned along the surface of the sample being studied. The AFM measures the repulsion between the electrons in the particles of the probe and the electrons in the particles of the sample surface. As the probe moves directly over a particle, it feels the stronger repulsive forces and the cantilever is bent upward. As the probe moves over the space between two particles, the repulsive force decreases and the cantilever relaxes back downward. While this is happening, a laser is bounced off the back of the cantilever onto a detector. Depending on where the laser strikes the detector, we can determine how much the cantilever is moving up and down. The resulting AFM image is similar to a topographic map showing how much the cantilever moves up and down as it scans over the surface of the sample.

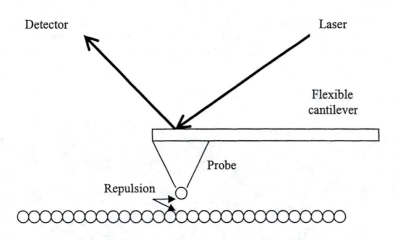

Cross section of surface of material being studied.
Circles represent individual submicroscopic particles.

Continued on back →

Photographs of the Setup of an Atomic Force Microscope (AFM)

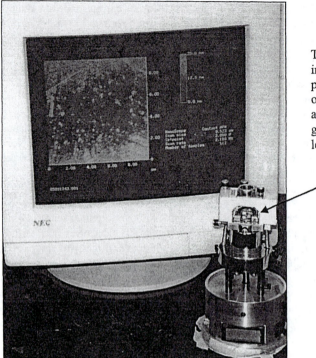

The photograph to the left shows the AFM instrument. The sample being studied and the probe/cantilever are housed in the small opening at the top of the AFM (indicated by the arrow). The image of the surface of a sample of gold (at the microscopic, not submicroscopic, level) is shown on the computer screen.

During the AFM experiment, the AFM instrument is placed in a plexiglass box and hung from a tripod by bungee chords (show, in the photograph below). The reason for this strange setup is that the AFM must be vibrationally isolated from the room around it. When you are trying to control the precise movements of the AFM probe, the minute vibrations caused by people walking around and closing doors is enough to ruin the image.

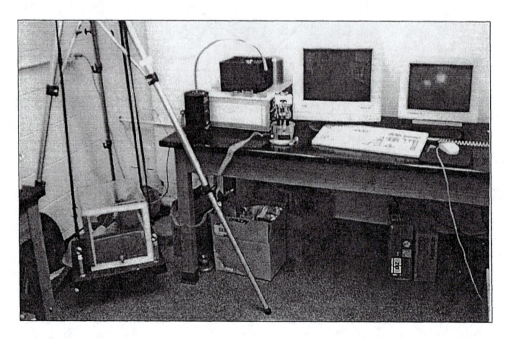

Explorations in Conceptual Chemistry: Activity 1b

Instructions for Model #1: Find the <u>foam-rubber packing material</u> and the <u>ruler with a die taped to the end</u>.

Questions:

1) How could these materials be used to model how an AFM works?

2) What objects in your model correspond to each part of the AFM?

3) How successful do you think the model is at conveying how AFM works? Your explanation should include a discussion of any limitations you see in the model.

4) Do the foam rubber and ruler/die serve as a conceptual or physical model for how AFM works? Explain.

Continued on Back →

Instructions for Model #2: Find the <u>flat refrigerator magnet</u> with the <u>strip cut off</u> the side.

Questions:
1) How could these materials be used to model how an AFM works?

2) What objects in your model correspond to each part of the AFM?

3) How successful do you think the model is at conveying how AFM works? Your explanation should include a discussion of any limitations you see in the model.

Instructions: Look at the image of a sample of graphite (carbon) below. The image was taken using a technique closely related to AFM called scanning tunneling microscopy (STM). While you are looking at the photo, think back to the article on AFM. Imagine the probe and cantilever scanning the surface of the sample as each line of the image is being made.

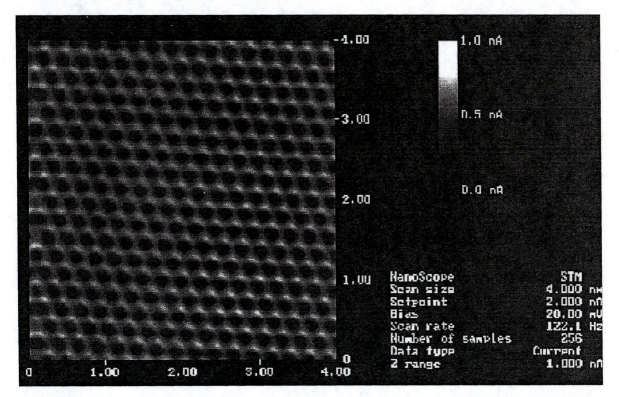

Questions:

1) Use a collection of <u>coins</u> to generate a representation of the arrangement of submicroscopic particles on the surface of carbon. In the space below, write some comments about what things strike you when looking at the STM image above and your representation using coins.

2) The scale on the image indicates that the length of the *x*-axis and the *y*-axis each is 4.00 nm. Knowing that 4.00 nm is equal to 0.000000004 meters, use the STM image to determine the radius, in meters, of a single submicroscopic particle of carbon. Show your work.

Name: _____   **Score:** _____ **points**

1) When we touch something that feels rough or bumpy, are we actually feeling the individual submicroscopic particles? If not, what are we feeling? As part of your answer, draw a submicroscopic cross section of a surface that you think would feel rough to the touch.

2) What is a topographic map? How is an AFM image like a topographic map?

Continued on Back →

3) As you observed in *Activity 1a*, the submicroscopic particles that make up matter are too small to see. In fact, the size that you calculated for the radius of a carbon atom is so small that we can't even begin to relate to it. It may help you understand how incredibly small these particles are if you consider that about 600,000,000,000,000,000,000 particles (that's a "6" with 20 zeros!) of aluminum could fit in the head of a common pin! To put this in perspective, perform the following calculation: "If every person on Earth (roughly 6 billion people) counted one submicroscopic aluminum particle every second, how many years would it take us to count the submicroscopic aluminum particles on the head of a pin?" Show all your work.

Unit 2: Phases, Phase Changes, and the Effect of Heat on Matter

In Unit 2, we really put our observational skills to work. While you may know about the three common phases of matter and the properties that we use to categorize something as a solid, liquid, or gas, you may not know what the submicroscopic particles are doing in each of the phases and what happens to them when a sample of matter undergoes a phase change, such as freezing or evaporation. To really understand the phases and phase changes requires that we further develop our model of the particulate nature of matter by making observations concerning the effect of temperature on the motion of submicroscopic particles. Along the way, we will take some time to learn about how gases exert pressure and about the Kelvin temperature scale. By the end of the unit you, will be able to explain a variety of things, such as how barometers are used to help predict the weather and why water boils below 100°C on the top of a mountain.

2

Activity 2a: The Properties of Solids, Liquids, and Gases
COVER SHEET

Before Class
- Read the pages in your textbook dealing with "density" and "solids, liquids, and gases." Your instructor may assign specific pages for you to read.
- Complete the Pre-Activity Problem Set on the next page.

Key Concepts
- Solid, liquid, and gaseous states of matter have distinct characteristic properties.
- The states of matter (solid, liquid, and gas) depend on the motion of their submicroscopic particles. In solids, the submicroscopic particles are closely locked in place, often in well-ordered arrangements, and can only vibrate in position; in liquids, the particles have a disordered arrangement and can move past one another; in gases, the particles are free to move about independently, colliding with each other and their container.
- Gases are compressible because there is empty space between the submicroscopic particles. Liquids and solids cannot, in general, be compressed because the particles are already as close to one another as they can get. For this reason, typical gases are less dense than typical solids and liquids.
- All matter has mass. Yes, even gases!

Learning Objectives
- You will be able to list characteristic properties of solids, liquids, and gases.
- You will be able to explain the properties of solids, liquids, and gases in terms of the behavior of their submicroscopic particles.
- You will be able to use the properties of solids, liquids, and gases to explain everyday phenomena.
- You will be able to draw pictures that illustrate an understanding of solids, liquids, and gases on the submicroscopic level.

Name: _____   Score: _____ points

1) List three examples each of a solid, a liquid, and a gas:
   Solid:

   Liquid:

   Gas:

2) You have volunteered to assist in "Science Day" at the local middle school. As part of your activity, you have decided that it would be fun at recess to have the students pretend they are submicroscopic particles of a solid, a liquid, and a gas. In the space below, write the directions that you would give the class to act out the behavior of the particles in each of the phases.

   Directions for a solid:

   Directions for a liquid:

   Directions for a gas:

3) What is meant by the term *buoyancy*? How can buoyancy change the apparent weight of an object?

Continued on Back →

4) Gallium has a melting point of 29.8°C and a boiling point of 2204°C.

    a)   Is gallium a solid, liquid, or gas at room temperature (around 25°C)? Explain.

    b)   In what state would gallium be if you held it in your hand (body temperature is around 37°C)? Explain.

CAUTION: Dry ice can cause severe burns. To avoid direct contact with exposed skin always use metal tongs and temperature-resistant gloves when handling dry ice.

Instructions: You will use <u>a flask with tapered walls (or a graduated cylinder or vase)</u>, a beaker (or measuring cup), several <u>small solids</u> (for example, beans, dice, rocks, etc.), a <u>funnel</u>, <u>water</u>, and <u>dry ice</u> to illustrate some of the key properties of solids, liquids, and gases. Begin by transferring each substance from the flask to the beaker. Then try to pour each substance through a funnel. When you are done, empty any remaining water/dry ice into the sink.

Observations:

Solid (Small Object)	Liquid (100 mL of Water)	Gas (See Note Below [**])
Transfer the solid from the flask to the beaker.	*Transfer the liquid from the flask to the beaker.*	*Transfer the gas from the flask to the beaker.*
What happens to the shape of the object?	What happens to the shape?	What happens to the shape?
To the volume of the object?	To the volume? [##]	To the volume?
What happens when you put the solid object into the funnel?	What happens when you put the liquid into the funnel?	What happens when you put the gas into the funnel? [++]

[**] Most common gases are colorless and the ones that aren't are toxic, so we will use the white fog that is formed when dry ice is added to water as a model for the behavior of a gas. Add about 200 mL of water to the flask and drop a piece of dry ice the size of the end of your thumb into the water (see **CAUTION** above). A white fog should be produced. **When pouring or transferring the gas, be sure to leave the water behind in the flask.**

[##] Ignore the volume markings on the beaker and flask. They are rough and not meant for accurate measuring.

[++] Make sure the funnel is totally dry and that there are no drops of water clogging the tip.

Instructions: Press the plunger on the <u>three syringes</u> that have been filled with a solid, a liquid, and a gas and then sealed off. Do NOT pull on the plunger since that can introduce air bubbles into the syringe. Observe what happens and record the results below.

Observations:

Solid-Filled Syringe	Liquid-Filled Syringe	Gas-Filled Syringe

Questions:

1) Explain the observations above in terms of the behavior of the submicroscopic particles.

   <u>Solid:</u>

   <u>Liquid:</u>

   <u>Gas:</u>

2) Make two labeled drawings: one of a sealed syringe filled with a solid and one of a sealed syringe filled with a gas. Your drawings should clearly indicate the behavior of the submicroscopic particles (represented by circles) in a way that explains your observations above.

3) What did you notice when you stopped pressing on the syringe plunger? What does this indicate that the gas particles are doing? Explain.

CAUTION: Dry ice (frozen carbon dioxide) can cause severe burns. Always wear insulated gloves when handling dry ice to avoid direct contact with exposed skin.

Instructions: Place one pellet of <u>dry ice</u> (or one level spoonful of crushed dry ice) into a <u>resealable plastic bag</u> and seal the bag. Place the baggie on the lab bench and record your observations for the next 5 – 6 minutes. Alternatively, crushed dry ice can be added to a balloon using a funnel. Tie off the balloon and record your observations.

Observations:

Questions:
1) The following questions seek to determine the relative densities of solids and gases.
   a) What is the mathematical formula that is used to determine density?

   b) What happened to the volume of the contents of the plastic bag (i.e., the carbon dioxide) during the phase change? Briefly defend your answer.

   c) What happened to the mass of the contents of the plastic bag (i.e., the carbon dioxide) during the phase change? Briefly defend your answer. Hints: What is responsible for the mass of the carbon dioxide? Does this change during the phase change?

   d) In which state is a sample of matter expected to be more dense (as a solid or as a gas)? Explain using your answers to questions a – c above.

CAUTION: Wear safety goggles during this activity to protect your eyes in case the bottle breaks due to the buildup of pressure.

Instructions: Add <u>water</u> to a <u>small, thick-walled plastic bottle</u> until it is ¼ <u>full</u>. Add an <u>Alka-Seltzer tablet</u> to the bottle and quickly close the cap tightly shut. Record the mass. Slowly loosen the cap on the bottle, so that no water bubbles come out when the gas escapes. Record the final mass.

Observations:

Mass of bottle, water, cap, and Alka-Seltzer (after adding the Alka-Seltzer to the water, closed bottle cap)	
Mass of bottle, water, cap, and Alka-Seltzer (after adding the Alka-Seltzer to the water, open bottle cap)	

Questions:

1) Most people would agree that all solids and liquids have mass. Why do you think it is so hard to imagine that gases have mass? Can you think of some practical examples where gases don't seem to have mass? What is it about the examples that you gave that make it seem like gases don't have mass?

2) How does this activity with the Alka-Seltzer help illustrate that gases have mass?

Instructions: Your instructor will have a student use <u>pennies</u> and the <u>overhead projector</u> to create submicroscopic scale representations of the particles in a solid, a liquid, and a gas. As the class critiques these representations, use the space below to take notes.

Questions:

1) Imagine the boxes below are containers holding solid, liquid, and gaseous samples of matter. Draw submicroscopic scale representations (with circles representing the particles) of the three samples. Your drawings should convey as much understanding as possible about the volume, shape, density, and compressibility of the three phases. In the space below the boxes, explain how each drawing relates to the observations you have made in this activity.

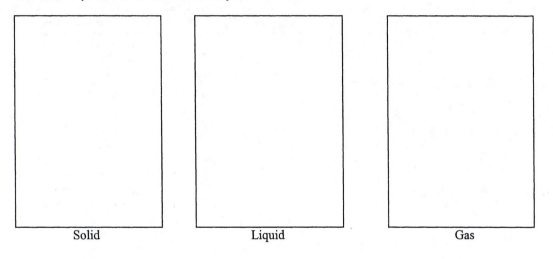

| Solid | Liquid | Gas |

Drawing of Solid:

Drawing of Liquid:

Drawing of Gas:

Continued on Back →

2) Complete the table below. Do the volumes and shapes of solids, liquids, and gases depend on their container? Which of the phases are compressible? Which expand to fill their container? Which have mass? Which have relatively low density?

	Solid		Liquid		Gas	
Does the volume depend on the container?	yes	no	yes	no	yes	no
Does the shape depend on the container?	yes	no	yes	no	yes	no
Can it be compressed?	yes	no	yes	no	yes	no
Does it expand to fill its container?	yes	no	yes	no	yes	no
Does it have mass?	yes	no	yes	no	yes	no
Does it have relatively low density?	yes	no	yes	no	yes	no

Explorations in Conceptual Chemistry: Activity 2a

Name: _____   Score: _____ points

1) In this activity, we saw that solids keep their own shape regardless of their container. Yet, when we add sand to a jar, the sand fills the bottom of the jar and takes the shape of the container like a liquid. Is the sand really a liquid? Explain what is happening in this case.

2) You and your lab partner are studying for chemistry and your lab partner goes into the kitchen to make a "surprise" snack. Within a minute, however, you detect a familiar smell from the other room and you know the snack is popcorn. What state of matter (solid, liquid, or gas) are you detecting with your sense of smell? What is the crucial property of this state of matter that allows you to smell it from the other room and that distinguishes it from the other two states? Explain. Draw a picture showing what is occurring to accompany your written explanation.

Continued on Back →

3) Look back at your results for *Part C,* "The Relative Densities of Solids, Liquids, and Gases." Draw two pictures of the bag (macroscopic scale is fine; in other words, you don't need to show the particles that make up the bag) with a submicroscopic representation of the dry ice inside (here you need to show the particles). One drawing should relate to your observation of the bag before the solid dry ice became a gas and one drawing of the bag after the dry ice has turned into a gas.

4) Some substances such as shaving cream, hair mousse, and canned whipped cream seem to have properties of both solids (they can keep their shape) and gases (they can expand to take up more space than their original containers). Are these substances a different state of matter or are they somehow a combination of the more common states (solid, liquid, and gas)? Explain your answer. Hint: Check out how an aerosol can works at http://science.howstuffworks.com/aerosol-can.htm

Activity 2b: Heat and the Motion of Submicroscopic Particles
COVER SHEET

Before Class

- Read the pages in your textbook dealing with "definition of heat and temperature" and "kinetic energy of particles." Your instructor may assign specific pages for you to read.
- Complete the Pre-Activity Problem Set on the next page.

Key Concepts

- The particles that make up matter are in constant motion. Even in solids the particles are vibrating.
- The temperature of a sample of matter is related to the speed of its submicroscopic particles. "Hot" samples have particles that are, on average, moving faster and "cold" samples have particles that are, on average, moving slowly.
- If the pressure is kept constant, the volume occupied by a gas increases as the temperature of the gas increases. If, however, the volume is not allowed to change, then the pressure that the gas exerts will increase as the temperature increases.
- Heat flows in solids by conduction, which involves no flow of matter (in other words, the particles vibrate in place and can collide with their neighbors, but they do not move from one place to another). Heat flows in gases and liquids by both conduction and convection, which involves the flow of matter (in other words, the particles can move from one location to another in the material). Both conduction and convection involve collisions between fast ("hot") particles and slow ("cold") particles. After the collision, the fast particles have lost energy and are now slower, and the slow particles have gained energy and have sped up.
- Though it can change forms (for example, kinetic to potential or thermal to electrical), energy can be neither created nor destroyed. This principal is called the law of conservation of energy.

Learning Objectives

- You will be able to relate the temperature of a sample to the motion of the submicroscopic particles that make up the sample.
- You will be able to discuss how changing the temperature affects a sample of gas.
- You will be able to explain how heat flows within an object and between objects.

Pre-Activity Problem Set: Activity 2b

Name: _____   **Score:** _____ **points**

1) When we heat water to make a cup of tea, what is happening to the particles that make up the water?

2) According to kinetic-molecular theory, all submicroscopic particles, even those of a solid, are always in motion. When we hold a solid in our hand, can we *directly* feel the individual particles moving? Can we *indirectly* feel the particles moving when we hold a solid? As part of your explanation, imagine your are holding a hot penny in one hand and a cold penny in the other hand.

3) The figure below represents the moving particles of a gas within a rigid container. Each circle represents a single submicroscopic gas particle. Of course, real particles DO NOT have tails behind them. The tails are an attempt to indicate the relative motion of the particles. In the space to the right of the figure, redraw the diagram as it would appear after the container was heated. Briefly justify/explain your drawing.

Original Temperature: Redrawn After Heating:

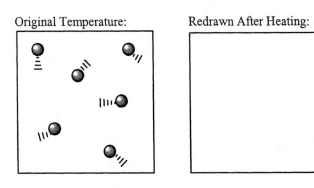

Part A: The Addition of Food Coloring to Hot and Cold Water

CAUTION: Hot plates can cause severe burns even long after they are turned off. Keep flammable objects (including your hands, clothes, and hair) away.

Instructions: Begin by reading through the rest of the instructions and then record your prediction in the space below. After making your prediction, place <u>two 100-mL beakers</u> on a sheet of <u>white paper</u>; add roughly <u>80 mL of hot water</u> to one beaker and <u>80 mL of cold water</u> to the other. Be sure to <u>wear gloves that provide heat protection</u> when handling the hot water. **Without stirring**, add 1 drop of <u>red food coloring</u> to each beaker and record your observations (using both words and drawings). When you are done, clean up your material according to your instructor's directions.

Prediction: Using any experiences you have had outside of class, what do you think will happen when you add the food coloring to the hot and to the cold water? On what did you base your prediction?

Observation:

Question:
1) Without using technical terms, provide a written description of what is happening on the submicroscopic level that will explain your observations in the two cases (hot versus cold water).

CAUTION: Hot water and glassware can cause severe burns. Always use <u>tongs</u> or wear <u>heat-resistant gloves</u> when handling hot objects.

Instructions: Blow up a <u>balloon</u> in a <u>medium-size beaker or glass</u> so that about half of the balloon is in the beaker and the other half is sticking out of the top. Tie off the end of the balloon. Stand the beaker and the balloon in a <u>container that is ½ full of hot water</u> (almost boiling) for about 5 minutes. Immediately transfer the beaker and the balloon to a <u>container of ice-cold water</u>. Record your observations (using both words and drawings).

Observations:

Question:
1) Draw a cutaway view of a balloon at high and low temperatures. Your drawing should use circles to represent the submicroscopic particles. Your picture should convey as much information as possible about what happens to the particles in the balloon and what happens to the size of the balloon at the two temperatures. Use arrows to indicate the relative pressure of the gases pushing outward and inward on the wall of the balloon.

- -

Optional Demonstration: Your instructor will demonstrate what happens when a <u>balloon</u> is placed in <u>liquid nitrogen</u>. Note the safety precautions your instructor takes when dealing with liquid nitrogen, which has a boiling point of -196°C. At that temperature, it would only take a few seconds to freeze your finger solid. Brrrr!

Observations:

Question:
1) Did the results of this demonstration agree with what you expected based on the behavior of the balloon in the hot and the cold water? Explain.

CAUTION: Hot plates can cause severe burns even long after they are turned off. Keep flammable objects (including your hands, clothes, and hair) away.

Instructions: You will find a <u>large beaker</u> filled with <u>sand</u> on a <u>hot plate</u> and a <u>beaker of water</u> with several <u>glass and aluminum rods</u>. Begin by reading through the rest of the instructions. When you are done recording your prediction, choose a cool glass and aluminum rod, and dry them off with a <u>paper towel</u>. Simultaneously, insert the rods all the way down in the sand. Let the rods remain in the sand for several minutes. Occasionally check the ends of the items sticking up above the sand to see if they are getting warm. When you are done, place the rods in the beaker of water to cool off.

Prediction: Using any experiences you have had outside of class, which rod do you think will get hot first? On what did you base your prediction?

Observations:

Question:
1) Employing our model of the particulate nature of matter, explain what is occurring on the submicroscopic level as heat travels the length of the metal rod.

2) Propose an activity that ten students could act out in order to help us visualize how heat is transferred along the length of a solid object (like a metal rod).

3) Propose a possible difference between the submicroscopic particles making up the glass rod and those making up the metal rod that could explain the observed differences in their ability to conduct heat.

CAUTION: Hot plates can cause severe burns even long after they are turned off. Keep flammable objects (including your hands, clothes, and hair) away.

Instructions: Take two <u>250-mL beakers</u>. Add roughly <u>100 mL of hot water</u> to one beaker and <u>100 mL of cold water</u> to the other. Be sure to wear <u>gloves that provide heat protection</u> when handling the large beaker of hot water. Take the beakers back to your desk. Use a <u>thermometer</u> to record the temperatures of the hot and cold water. Record your prediction of the temperature after the hot and cold water are mixed. While stirring with a <u>glass rod</u>, add the cold water to the hot water and again record the temperature.

Prediction:
Temperature (°C) of the mixed hot and cold water:

Observations:

	Hot Water	Cold Water	Mixed Water
Actual Temperature (°C)			

Question:
1) What must happen between the particles of the hot water and the cold water if there is to be a transfer of energy between them?

2) Does the final temperature of the water make sense? Explain your answer in terms of what has happened to the speed of the submicroscopic particles in the hot water when they came in contact with the submicroscopic particles making up the cold water. Your answer should make reference to the law of conservation of energy (see "Key Concepts" on page 41 of this activity).

3) What do you predict would be the final temperature if instead you added the 100 mL of cold water to 200 mL of the hot water? Fully explain your answer in terms of the behavior and relative numbers of the faster-moving and slower-moving submicroscopic particles.

Name: **Score:**   **points**

1) The following questions refer to the terms *conduction* and *convection* as described under the "Key Concepts" on the cover sheet of this activity (page 41).

   a) Which part of this activity clearly represents a case of heat flow by conduction? Explain how your observation fits the definition of conduction.

   b) Which part of this activity clearly represents a case of heat flow by convection? Explain how your observation fits the definition of convection.

   c) Following the example back in Question #2 of *Part C*, propose an activity that ten students could act out that would help us visualize how heat is transferred throughout a gaseous object (imagine a fireplace in one corner of a room that heats the entire room).

2) When a hot piece of metal is dropped into a pail of room-temperature water, what happens to the temperature of the water? What happens to the temperature of the metal? Your explanation should include a discussion of the transfer of heat, the collisions between the submicroscopic particles, and the law of conservation of energy.

Continued on Back →

3) A thermos can keep hot things hot and cold things cold. How does it know which to do? Given the following diagram of a thermos and knowing that in a vacuum there are very few gas particles moving around, explain how you think a thermos works. Again, your explanation should include a discussion of the transfer of heat, the collisions between the submicroscopic particles, and the law of conservation of energy.

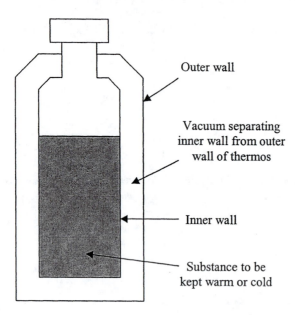

Outer wall

Vacuum separating inner wall from outer wall of thermos

Inner wall

Substance to be kept warm or cold

Activity 2c: Understanding Gas Pressure
COVER SHEET

Before Class
- Read the pages in your textbook dealing with "gases" and "atmospheric pressure." Your instructor may assign specific pages for you to read.
- Complete the Pre-Activity Problem Set on the next page.

Key Concepts
- Gases exert pressure on objects by collision between the submicroscopic gas particles and the object.
- Gas particles move from regions of high pressure to regions of low pressure.
- Atmospheric pressure decreases with increasing altitude.
- Atmospheric pressure plays a crucial role in weather patterns and in everyday life.

Learning Objectives
- You will be able to discuss how gases exert pressure.
- You will be able to explain in words and drawings everyday situations involving atmospheric pressure.
- You will be able to make and use a simple barometer for measuring changes in atmospheric pressure.

Name: _____   Score: _____   points

1) Imagine you are standing about 2 feet away from a fan set on high speed. Describe what is happening and what you are feeling on your face in terms of submicroscopic gas particles.

2) The two boxes below represent rigid, sealed containers holding a sample of gas. Using the example in Box #1, complete the drawing in Box #2, showing what would happen if gas particles are added to Box #1. In the space below the Box #2, explain which of the two boxes would have the higher pressure and why. Make sure you come up with a way to use the arrows to indicate which box has the higher pressure.

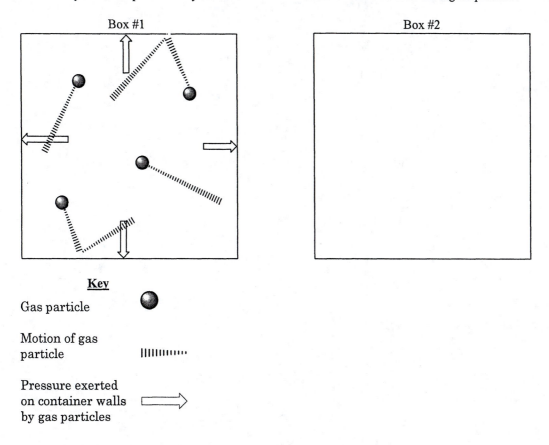

Box #1   Box #2

Key

Gas particle

Motion of gas particle

Pressure exerted on container walls by gas particles

Continued on Back →

3) One way to report pressure is pounds per square inch (lb/in^2). These units may seem strange at first, but they make sense if we think about it. Explain.

4) Other than lb/in^2, what are three other units for pressure? What units are usually used when reporting atmospheric/barometric pressure during the weather forecast?

Instructions: Squash a <u>paper towel</u> in the bottom of a <u>small beaker</u>. The paper towel should only take up the bottom half of the beaker and should stay in place when the beaker is flipped upside down. Submerge the inverted beaker and paper towel in a <u>large beaker</u> half-filled with <u>water</u> so that the paper towel is clearly below the level of the water. Remove the inverted beaker and examine the paper towel.

Observations:

Question:
1) Draw and label a picture that explains, on a submicroscopic level, your observations. Use arrows to indicate the pressure exerted by the gas particles.

Alternate Instructions: Insert 1 – 2 inches of a <u>straw</u> into a <u>plastic baggie</u>. Hold the baggie shut around the straw so that no air can escape. Place your chemistry <u>book</u> on top of the baggie. Use the straw to blow up the baggie. Record your observations.

Observations:

Question:
1) Draw and label a picture that explains, on a submicroscopic level, your observations. Use arrows to indicate the pressure exerted by the gas particles.

CAUTION: Be careful when using the scissors to make holes in the bottle. Your instructor may already have prepared the bottle with holes for you.

Instructions: Use a <u>pair of scissors</u> to punch three holes along one side (one each near the top, the middle, and the bottom) in a <u>2-liter bottle</u>. Cover the holes with a single strip of <u>tape</u> and fill the bottle with <u>water</u>. Place the bottle in the <u>sink or in a large bucket or pan</u>. Remove the strip of tape and record your observations in the form of a labeled drawing.

Observations (Labeled Drawing):

Questions:

1) Is the pressure on the water greatest at the top or at the bottom of the bottle? Support your answer using your observations.

2) Explain the cause of the variation in pressure that you observed in Question #1.

3) How is the atmosphere like a deep pile of feathers? Fully explain what happens at the top of the atmosphere (the feathers) and the bottom of the atmosphere (the feathers) and why.

4) Why might you expect a baseball to fly farther when thrown at high altitude (as in Denver) versus low altitude (as in San Diego)?

CAUTION: Be careful when using the scissors to make holes in the bottle. Your instructor may already have prepared the bottle with holes for you.

Instructions: Use a <u>pair of scissors</u> to punch a small hole in the bottom of a <u>2-liter soda bottle</u>. Place the body of a <u>balloon</u> inside the bottle and stretch the opening of the balloon around the entire mouth of the bottle. Record your observations as you try to blow up the balloon while covering the hole in the bottom. Next, uncover the hole and again try to blow up the balloon. When the balloon is full, put your finger back over the hole and take your mouth off the balloon. Finally remove your finger from over the hole.

Observations:

Questions:
1) Why couldn't you blow up the balloon when the hole was plugged?

2) After initially blowing up the balloon with the hole open, why did covering the hole keep the balloon inflated?

Alternate Instructions: Your instructor has prepared a <u>small balloon sealed in a syringe</u>. Alternate between pushing and slowly pulling on the plunger (being careful not to completely pull the plunger out) and observe what happens to the balloon.

Observations:

Question:
1) Explain in terms of submicroscopic gas particles what is happening to the small balloon when the plunger is pulled out.

Instructions #1: To make a simple **homemade barometer** all you need is a <u>clear drinking straw</u>, and a <u>glass</u> of <u>water</u>. Draw water about half-way up the straw. Trap the air in the top half of the straw by sealing the top of the straw (fold over about ½ inch of the straw and <u>tape</u> it down). Carefully mark the level of the water in the straw. To use your barometer, place the straw back in the half-filled glass of water. If you don't have several days to wait to see changes in atmospheric pressure, just leave your barometer near a <u>large beaker</u> of <u>water</u> left on a <u>hot plate</u> to boil for 10 – 15 minutes. Make sure the barometer doesn't heat up from the hot plate by shielding it with a textbook.

Observations:

Questions:

1) Explain how your barometer works. Include a labeled drawing as part of your answer.

2) What happened to the level of the water in your barometer when it was placed near the boiling water? Does this result suggest that moist air is more or less dense than dry air? Briefly explain.

3) Do your observations make sense in terms of weather forecasting that says when a low-pressure area is moving toward your region, this usually means cloudy weather and precipitation are on the way? Explain.

Continued on Next Page →

Instructions #2: To build a **working model of a lung**, begin with a <u>clear plastic baby bottle</u> that has had its bottom removed. Use <u>scissors</u> to cut the neck off a <u>balloon</u>. Secure the balloon in place over the open bottom of the baby bottle with a <u>rubber band</u>. Cut a <u>3-inch piece of straw</u> and put it through the hole in a <u>rubber test-tube stopper</u> so there is 1 inch sticking out of the top. If necessary, secure the straw in place with <u>glue or tape</u>. Attach a <u>second balloon</u> to the piece of the straw sticking out of the bottom of the rubber stopper, using glue or tape to secure it in place as needed. Tightly insert the rubber stopper with the balloon into the baby bottle. Record your observations as you alternate between pulling down and releasing the bottom balloon.

Observations:

Questions:

1) What happens when you pull on the bottom balloon? Why does this happen?

2) How does the baby bottle and balloon serve as a model for how the lungs work? As part of your answer, discuss how each part of your model corresponds to each part of the lung.

Name: **Score:**   **points**

1) Imagine you drink half a bottle of water on a camping trip to the mountains. After putting the cap back on the bottle, you pack it in your car and drive back down to your home in the valley. At home, you notice that the bottle is slightly squashed and that it returns to its normal size when opened. Following the example drawn for the closed bottle in the mountains, finish making labeled drawings to clearly explain what is happening to the bottle after it is brought down to the valley. In the spaces below the last two drawings, briefly explain what is happening.

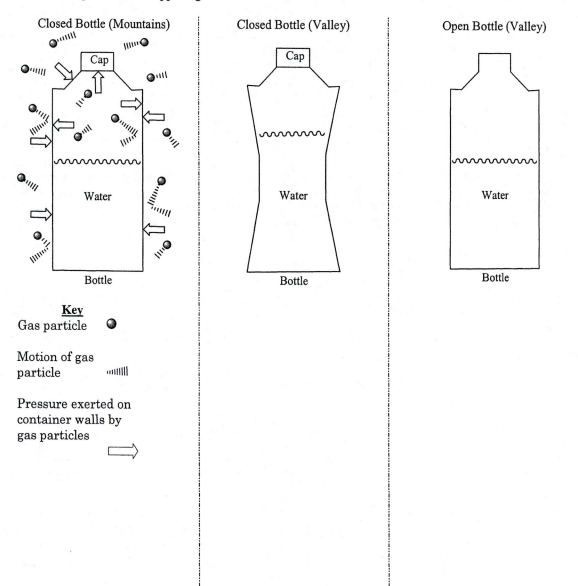

Closed Bottle (Mountains)   Closed Bottle (Valley)   Open Bottle (Valley)

Key

Gas particle

Motion of gas particle

Pressure exerted on container walls by gas particles

Continued on Back →

2) Most people think that when they use a straw, they are sucking the drink into their mouths. A better way to understand how a straw works is to consider that the atmosphere is actually **pushing** the liquid up the straw. Using what you have learned about the pressure exerted by gases in the atmosphere, explain how a straw works on a submicroscopic level.

3) You bought a dozen helium balloons for your friend's birthday and left them in your car while you were in school. The hot sun beat down on your car all day (it was August) and when you went back to you car at the end of the day, all the balloons had popped. Explain what happened to the balloons in terms of the speed and pressure exerted by the submicroscopic gas particles inside of the balloons.

Activity 2d: Absolute Zero and the Kelvin Temperature Scale
COVER SHEET

Before Class
- Read the pages in your textbook dealing with "absolute zero" and the "Celsius, Fahrenheit, and Kelvin temperature scales." Your instructor may assign specific pages for you to read.
- Complete the Pre-Activity Problem Set on the next page.

Key Concepts
- The Kelvin temperature scale (or absolute-zero scale) is based on the theoretical temperature limit where a sample of gas has no volume.
- Graphical representations of data often allow scientists to more easily draw conclusions about the data.

Learning Objectives
- You will be able to explain the difference between heat and temperature.
- You will be able to describe the basis for and the significance of the Kelvin temperature scale.
- You will be able to demonstrate an understanding of temperature scales by creating a new temperature scale.
- You will be able to construct a graph. You will then be able to use the graph to draw conclusions.

Name: _____ Score: _____ points

1) What is the difference between temperature and heat?

2) What physical standards are the Fahrenheit and Celsius temperature scales linked to?

   Fahrenheit Scale:

   Celsius Scale:

3) For a temperature scale to be useful, it should be easily reproduced by other people who would like to be able to make a thermometer using that scale. Why are the physical standards to which we link the Celsius scale such a good choice? Briefly explain how you might go about making a thermometer using the Celsius scale.

4) What is the formula for converting from °F to °C? Water is most dense at 4°C. What is this temperature in °F?

The following table provides experimental data for the volume of a helium balloon at various temperatures.

Temperature (°C)	Volume (L)
100	3.77
24	3.00
-196	0.78

1) Why do you think the scientists chose to make measurements at each of the above temperatures? (In other words, what could a scientist do in a laboratory to a balloon to get it to that temperature?)

Plot the above data on the template, "The Effect of Temperature on the Volume of a Helium Balloon." Draw the best-fit line between the data points. Extrapolate the data so that the line passes through the x-axis.

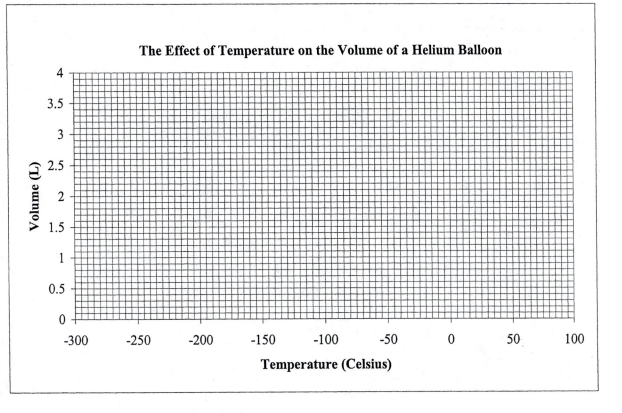

The Effect of Temperature on the Volume of a Helium Balloon

Continued on Back →

2) At what temperature does the line pass through the x-axis? What is the significance of the volume at this temperature?

3) Use the graph on the previous page to determine the volume of the balloon at 0°C. Clearly explain the steps for how you came up with your answer.

The Kelvin scale is made by assigning 0 K to where our line crosses the x-axis. We then add 273 to all the other temperatures on our graph. Rewrite the temperatures on your x-axis in terms of absolute-temperature (Kelvin) scale.

4) Which temperatures on your x-axis cannot be rewritten in the Kelvin scale? Explain.

5) What are the melting point and boiling point of water on the Kelvin scale?

6) List four things that the template has that are needed for making a good graph.

Explorations in Conceptual Chemistry: Activity 2d

Name: **Score:**   **points**

1) Look back at the Pre-Activity Problem Set to see the physical standards used to develop the Celsius and Fahrenheit temperature scales. Now, come up with a new temperature scale that is based on different physical standards. Be as imaginative as possible.

2) A student suggests making a temperature scale based on the temperature at which a Hershey's chocolate bar begins to melt. A quick experiment reveals that this occurs at 30°C. This temperature will therefore become zero degrees on the Hershey scale (0°H). In addition to designating a new zero, the student decided that she did not need her temperature scale to be very precise, so a temperature increase of 10°C will correspond to a temperature increase of only 1°H.

   a) Using this information, relabel the Celsius temperature scale below using the Hershey temperature scale.

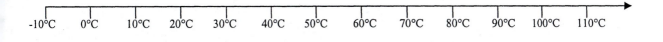

-10°C   0°C   10°C   20°C   30°C   40°C   50°C   60°C   70°C   80°C   90°C   100°C   110°C

   b) At what temperature does water freeze on the Hershey scale?

   c) At what temperature does water boil on the Hershey scale?

3) Which has a higher temperature, a swimming pool of boiling water or a coffee cup of boiling water? Which has more heat? Fully explain your answer.

Activity 2e: Exploring the Phase Changes
COVER SHEET

Before Class
- Read the pages in your textbook dealing with "phase changes." Your instructor may assign specific pages for you to read.
- Complete the Pre-Activity Problem Set on the next page.

Key Concepts
- Phase changes are accompanied by the input or release of heat.
- Phase changes are reversible.
- Phase changes are examples of physical changes because the same submicroscopic particles are present before and after the phase change. The difference before and after the phase change is the amount of kinetic energy that the particles have.
- Each substance has its own characteristic amount of stickiness between its particles. The amount of stickiness is related to the freezing and boiling points of that substance.

Learning Objectives
- You will be able to name all the phase changes.
- You will be able to draw submicroscopic representations of all the phase changes.
- You will be able to explain all the phase changes in terms of heat flow and the behavior of the submicroscopic particles.
- You will be able to relate phase changes to common everyday occurrences.
- You will be able to construct a simple model of the water cycle and identify the phase changes that are occurring.

Name: _____ **Score:** _____ **points**

1) Draw a line matching the description of each phase change with the corresponding name.

Description	Name
solid → liquid	melting
solid → gas	evaporation
liquid → gas	freezing
liquid → solid	sublimation
gas → solid	deposition
gas → liquid	condensation

2) For each of the phase changes, indicate whether the change requires an input or a release of heat (energy). In other words, if you wanted to perform each phase change, would you use a stove (to input energy) or a refrigerator (to remove energy)? Circle the appropriate word in the sentences below.

Melting requires a(n) <u>input/release</u> of heat (energy).

Evaporation requires a(n) <u>input/release</u> of heat (energy).

Freezing requires a(n) <u>input/release</u> of heat (energy).

Sublimation requires a(n) <u>input/release</u> of heat (energy).

Deposition requires a(n) <u>input/release</u> of heat (energy).

Condensation requires a(n) <u>input/release</u> of heat (energy).

3) On the back of this page, come up with a visual way to summarize the information in Questions #1 and #2. Your representation should also include submicroscopic drawings of the three states.

CAUTION: Always be careful when dealing with fire. Do not leave the lit candle unattended.

Instructions: Use a <u>match or lighter</u> to light a <u>small votive candle</u>. Record your observations for the next 2 – 3 minutes. When you are done, carefully blow out the candle. Again record your observations after 2 – 3 minutes.

Observations:

Questions:

1) What is the name of the phase change that the wax underwent? This phase change requires an input of heat. Where did the heat that sped up the wax particles come from?

2) Draw two pictures of the candle illustrating the submicroscopic particles of the wax before and after the phase change.

3) Name the phase change that happens to the wax after the candle is blown out. Does this phase change release heat or absorb heat? Where did the heat go to/come from?

4) Which requires more heat to become a liquid—wax or a rock? What does your answer suggest about the amount of "stickiness" holding together the submicroscopic particles in the wax versus those in the rock? Explain your answer in terms of the amount of heat that is needed to overcome the stickiness between the particles so that the phase change can occur.

CAUTION: Hot plates can cause severe burns. Use heat-resistant gloves to handle hot glassware.

Instructions: Set up your model of the water cycle as shown in the drawing below. Place a 1000-mL beaker on a hot plate. Add about 100 mL of water to which you have added a few drops of food coloring. Place an empty 100-mL beaker in the bottom of the large beaker. Cover the large beaker with plastic cling wrap so that there is a dip in the middle, just over the small beaker. Place several ice cubes in the center of the plastic cling wrap. Place the set-up on a hot plate turned to a medium setting. For best results, don't let the water come to a boil. Answer the questions below.

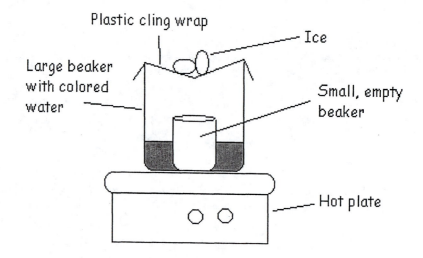

Questions:
1) All three states of water (solid, liquid, and gas) are present in your model. On the macroscopic-scale sketch below, draw and label submicroscopic representations of each of the states of water in the place where it is occurring.

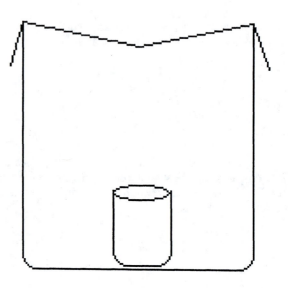

Continued on Next Page →

2) Where is evaporation occurring in your model? Why does it occur there?

3) Where is condensation occurring? Why does it occur there?

4) What do you notice about the water collecting in the small beaker? Explain.

5) If you were stranded on a desert island with no water to drink, how could you use the results from Question #4 to make safe drinking water from the ocean water?

6) How does our model relate to precipitation and the water cycle? To what does each part of our model correspond?

Part C: Return of the Disappearing Liquid

Instructions: Wet a <u>cotton ball</u> with <u>acetone</u> (<u>fingernail polish remover</u>). The cotton ball should be damp but not soaked. Wrap the damp cotton ball around the end of the <u>thermometer</u> and lay the thermometer and cotton ball on a <u>paper towel</u>. Record the temperature of the thermometer every minute for 5 minutes. You can gently blow on the cotton ball to help speed the process.

Observations:

	1 Min	2 Min	3 Min	4 Min	5 Min
Temperature (°C)					

Questions:

1) What phase change is the acetone undergoing in this experiment?

2) To undergo this phase change, the acetone particles must have sped up. According to the law of conservation of energy, if the acetone particles sped up, something else must have slowed down. What slowed down? In other words, what provides the energy to the acetone particles? Support your answer using your observations.

3) Explain how sweating can help to regulate our body temperature when it is hot outside or when we are taking part in physical exertion.

Name: _____   Score: _____ points

1) Write the name of the phase change in the blank next to the correct description.

   This phase change occurs when ice "disappears" in the freezer.   _____

   This phase change helps keep our bodies cool when we sweat.   _____

   This phase change occurs when heat is removed from a liquid.   _____

   This phase change occurs when your mirror fogs up after a hot shower. _____

2) Using the picture below (left) as a model, draw and label a single picture showing the submicroscopic particles (represented by circles) of a sample of matter undergoing any of the six phase changes other than the one already shown. Notice how the arrows are used to indicate that the particles start as a solid and end up in the gas phase. Below the drawings, provide a description of what is occurring in the first picture. Your description should include a discussion of the heat added (or removed), the kinetic energy (motion) of the submicroscopic particles, and whether the particles are sticking together more or less strongly.

   What phase change is shown in this   What phase change is shown in your
   drawing?   drawing?

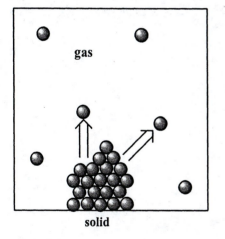

   Description of Phase Change Occurring Above (Left):

   Description of Phase Change Occurring Above (Right):

Continued on Back →

3) A cube of butter is placed on a piece of bread that has just come out of the toaster. Within 1 minute, the cube of butter has become a puddle. Using what we have learned so far this semester, explain what has happened to the butter.

4) The Huygens space probe, which has recently provided us with the first detailed glimpses of Saturn's moon, Titan, is indicating a bizarre atmosphere where methane rains down on the surface as a liquid. On Earth, methane is typically found as a gas. What conditions on Titan might explain why something that we normally think of as a gas might be found as a liquid. Support your answer using what you know about the behavior of submicroscopic particles in the gas and liquid phases.

Activity 2f: The Difference Between Boiling and Evaporation
COVER SHEET

Before Class
- Read the pages in your textbook dealing with "evaporation" and "boiling." Your instructor may assign specific pages for you to read.
- Complete the Pre-Activity Problem Set on the next page.

Key Concepts
- Water does not always boil at 100°C. For example, temperatures greater than 100°C are required to make water boil in a pressure cooker, whereas at high altitude, water boils at temperatures less than 100°C.
- Not all of the submicroscopic particles (SMP) in a sample of a matter are moving at the same speed. Some particles are moving slowly and some are moving very quickly. It is these fast particles that can enter the gas phase by overcoming the stickiness attracting it to its neighboring particles.
- The *average* speed of the particles in a sample of matter is related to the temperature of the sample. As the sample is heated, the particles have, on average, greater speed.
- Though both boiling and evaporation involve submicroscopic particles going from the liquid to gaseous states, boiling occurs when the vapor pressure of the liquid is equal to the atmospheric pressure.

Learning Objectives
- You will be able to explain the difference between evaporation and boiling.
- You will be able to explain everyday observations in terms of understanding what is occurring during evaporation and boiling.
- You will be able to discuss the factors that affect the boiling point of a liquid.

Name: Score: points

1) Remove a piece of blank graph paper from the back of this activity manual and prepare it as follows, so it will be ready to use when you do this activity. Hold the page horizontally so that the x-axis is on the long side of the page. Title the graph "Distribution of SMP Speeds in a Sample of Water (at High and Low Temperatures)." The x-axis should be called "Relative SMP Speed" and should have a range of 0 to 25. The y-axis should be labeled "Percent of SMPs" and should have a range from 0 to 20. The ranges on the x-axis and y-axis should take up as much of the page as possible. Attach your prepared graph.

2) In hot, dry environments the water in a swimming pool must be replenished regularly. What happened to the missing water?

3) Use the example in Question #2 to explain that there must be a difference between boiling and evaporation.

4) While boiling water to make pasta, you notice bubbles forming in the water and rising to the surface, where they pop. What is in the bubbles? Explain your answer.

Instructions: Table 1 shows the relative speed of the submicroscopic particles (SMPs) in a sample of water at different temperatures. Using the graph that you prepared as part of your Pre-Activity Problem Set, graph the data from the **first two columns** and connect the data points to form a smooth curve (you are **NOT** drawing a best-fit line in this case).

Table 1

Relative Speed	% of SMPs with That Speed (Low Temperature)	% of SMPs with That Speed (High Temperature)
0	0	0
1	5	2
2	11	4
3	16	7
4	18.5	10
5	17	13
6	14	15
8	8	17
10	5	14
12	3	10
15	1.5	4.5
17	1	2.5
20	0	1
25	0	0

Questions:

1) According to your graph, what is the most common relative speed (the one with the highest %) of a water particle at this low temperature? What is the range (highest and lowest value) of water particle speeds at this low temperature?

   Most Common Speed: Range of Speeds:

2) In a previous activity, we drew the figure below (left) to indicate the motion of particles in the gas phase. The imaginary tails were used to indicate the motion of the particles. Assuming the same type of thing is occurring with gases as with the liquid in Table 1, how might you revise this drawing to more accurately depict the motion of gas particles? Briefly explain your drawing.

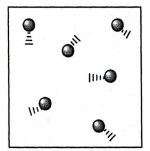

Drawing shows all gas particles
moving at the same speed.

Continued on Back →

3) An SMP in our sample of liquid water must have a minimum relative speed to overcome the attraction to its neighbors that keep it in the liquid state. Let's assume that a water particle requires a relative speed of "15" or greater to enter the gas phase. We can indicate this on our graph by drawing a vertical line at 15. What percentage of the water SMPs can evaporate at this temperature?

4) Even though a water particle *technically* has the minimum required speed, it still might not evaporate. What other factor do you think controls whether an SMP can move from the liquid to the gas phase? Hint: Where in the liquid does a particle have to be in order to evaporate?

5) Based on the previous question, would you expect a sample of water to evaporate more quickly in a shallow pan or in a tall, narrow vase? Explain.

Now, imagine that we heat our water sample and again measure the distribution of SMP speeds. Go back to your graph and add the high temperature data (the **third column** in Table 1). Make a second smooth curve with this new data (maybe in a different color so it will stand out) and answer the following questions.

6) Explain in words how your drawing of a gas in Question #2 would change after the sample has been heated to a higher temperature.

7) Even though the temperature has changed, a water SMP still needs the same relative speed to enter the gas phase (i.e., "15"). What happens to the % of particles that can enter the gas phase after the temperature is increased?

8) Use your answer to Question #7 to explain why you would expect water to evaporate faster from a bird bath on a hot summer day than on a cold winter day.

Explorations in Conceptual Chemistry: Activity 2f

Now let's look at the pressure due to the gases created above the water at the two temperatures discussed in *Part A*.

The drawing below (left) shows that at low temperatures, there is very little evaporation because most of the water particles (shaded circles) did not have a high enough speed to overcome their stickiness. The maximum pressure created by the amount of a liquid that has evaporated at a given temperature is called the **vapor pressure** (indicated by the small arrow). In this case, there are only a few gas particles, so the liquid is said to have a low vapor pressure. The large arrow outside the box represents the atmospheric pressure due to the air particles (solid black circles) in the atmosphere pushing down on the gas.

Low Temperature   **High Temperature**

Question:
1) Complete the drawing above (right) for a sample of water at a high temperature. Briefly defend your picture. What happened to the vapor pressure at high temperature?

As we continue to heat the liquid, the vapor pressure continues to increase (since more and more particles will have the required minimum speed to overcome their stickiness). When the vapor pressure of the gas being generated by evaporation is equal to the atmospheric pressure, then the liquid can finally boil. In other words, a liquid boils when it reaches the temperature at which there are enough particles that have become a gas (because they have the required speed) to generate a pressure that is equal to or greater than the atmospheric pressure.

In a sense, the average particle in the liquid is now moving so fast that the atmosphere is no longer able to push down and hold the liquid together. In the case of water, that temperature is usually 100°C. As we will see in *Part C*, however, if we can somehow change the "atmospheric pressure," we can make the water boil at a higher or lower temperature.

CAUTION: Take care when dealing with hot plates and hot glassware. Your instructor will use heat-resistant gloves when handling/inverting the Erlenmeyer flask.

Instructions: Your instructor will demonstrate how to boil water with ice. A 250-mL Erlenmeyer flask with 50 mL of water is left to start boiling on a hot plate. Once the water is boiling vigorously, the flask is sealed with a rubber stopper and inverted. Several ice cubes are placed on the flat bottom of the upside-down flask. Record your observations.

Observations:

Question:

1) Fully explain how the water is able to boil though it is no longer at 100°C. As part of your answer, complete the drawings below, showing submicroscopic particles of water in the liquid and gas phases before and after the ice is placed on the flask. Label your drawings to indicate the speed of the gas particles and the magnitude of the pressure created by the gas.

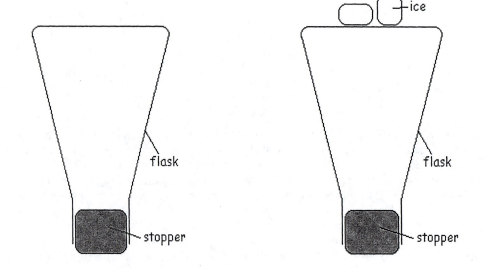

Name: _____   Score: _____ pts

1) Not all liquids boil at the same temperature. Some liquids, such as rubbing alcohol and gasoline, boil at a lower temperature than water. Answer the following questions about the boiling point of gasoline assuming that gasoline particles require a relative speed of "10" to become a gas, and also assuming that the gasoline has the same distribution of particle speeds shown in Table 1, *Part A*.

a) At a given temperature, do more gasoline particles or do more water particles have enough speed to evaporate? Explain.

b) Would you expect gasoline to have a higher or lower vapor pressure than water at a given temperature? Explain.

c) Explain why gasoline boils at a lower temperature than water. Be sure to discuss vapor pressure as part of your answer.

d) Care must be taken not to store liquids, such as gasoline, in tightly sealed containers in extreme heat. Explain why a gasoline container might explode at high temperatures.

Continued on Back →

2) When camping at the top of a mountain, it takes longer to hard boil an egg because the water boils at a temperature lower than 100°C. Fully explain why the water is able to boil at less than 100°C at high elevations. Include a labeled drawing (like the ones in *Part B*) with your answer.

Unit 3: An Overview of the Periodic Table

Without a doubt, the most important tool that a chemist has isn't a test tube or a hot plate... it is the periodic table. You've most likely seen that strange poster hanging in at least one classroom. You may even have heard that when you study chemistry you'll have to memorize all the symbols and numbers on the chart. Well, nothing could be further from the truth. The periodic table isn't something that you will be burdened with memorizing, rather it will be your trusty companion from here on out as you learn chemistry. After you learn how to dissect the richness of information the periodic table contains, you'll never want to be without it! In Unit 3, you'll follow in the footsteps of the Russian scientist Dmitri Mendeleev as you organize the properties of the elements to make your own periodic table.

Activity 3: Properties of the Elements and the Periodic Table
COVER SHEET

Before Class

- Read the pages in your textbook dealing with the "periodic table." Your instructor may assign specific pages for you to read.
- Complete the Pre-Activity Problem Set on the next page. You will need to use the Internet for some parts of the Pre-Activity Problem Set.
- Cut out the two pages of Element Property Cards found immediately after the Pre-Activity Problem Set. Be sure to cut out the blank cards too. Put them in an envelope for safekeeping and bring them to the lab.

Element A
• 12
• solid
• nonmetal
• 1:2 ratio with O
• 1:4 ratio with H
• no reaction (water)
• 3550°C
• dark powder

Key Concepts

- All matter is composed of various combinations of the 112 different types of atoms.
- Each element is made of one kind of atom. The 112 known elements are organized on the periodic table according to trends in their properties.
- Comparing early and modern versions of the periodic table can provide insight into the history and development of science.
- Living organisms and most materials are composed of just a few elements.

Learning Objectives

- You will be able to group common objects by their properties. You will be able to relate this type of organization to the organizational foundation of the periodic table.
- You will be able to use the periodic table to make predictions concerning the properties of the elements.
- You will be able to describe the historical basis by which the periodic table was developed.
- You will be able to compare and contrast early versions of the periodic table with the modern periodic table.
- You will be able to identify important groups on the periodic table such as the halogens, the transition metals, and the noble gases.
- You will be able to identify regions corresponding to metals, nonmetals, and inert gases.
- You will be able to write the names and symbols for the most common elements.

Name: _____   Score: _____   points

1) Using your textbook, write a definition that clearly distinguishes between the terms *atom* and *element*. Which term refers to matter on the submicroscopic level and which term refers to matter on the macroscopic level?

2) When a chemist writes "CS," is it the same as writing "Cs?" Explain.

3) Using an online periodic table (for example, http://www.webelements.com/), select an element and answer the following questions.
   a) What element did you choose (name and symbol)?

   b) What are some uses of the element you chose?

   c) What year was your element discovered/created?

4) The following questions refer to a typical square on the periodic table, for example, the one shown below to the right.
   a) What is the top number called?

   b) What does the top number indicate?

   c) What is the letter in the middle called?

   d) In this case, what does the letter represent?

   e) What is the bottom number called?

   f) What does the bottom number represent?

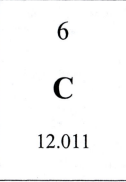

Continued on Back →

5) The following directions refer to the blank periodic table below.
   a) Label the following terms on the periodic table: alkali metals, alkaline earth metals, transition metals, inner transition metals, lanthanides, actinides, chalcogens, halogens, and noble gases.
   b) Number the periods and the groups on the periodic table.
   c) Color-code the portions of the periodic table where you'd find the metals, nonmetals, and metalloids.

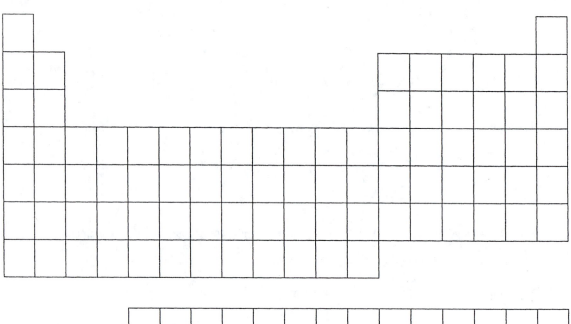

6) Answer the following questions.
   a) What is the symbol and the name of the element in group 16, period 5?

   b) What is the symbol and the name of the alkaline earth metal in period 6?

   c) What is the symbol and the name of the only nonmetal in group 14?

7) Write a question of the type found in Question #6 above. Provide an answer to the question.

Element Property Cards (Page 1)

Element A	Element B	Element C
• 12	• 128	• 39
• Solid	• Solid	• Solid
• Nonmetal	• Metalloid	• Metal
• 1:2 ratio with O	• 1:3 ratio with O	• 2:1 ratio with O
• 1:4 ratio with H	• 1:2 ratio with H	• 1:1 ratio with H
• No reaction (water)	• No reaction (water)	• Violent reaction (water)
• 3527°C	• 450°C	• 63°C
• Black	• Silvery gray	• Silvery white/soft

Element D	Element E	Element F
• 80	• 127	• 20
• Liquid	• Solid	• Gas
• Nonmetal	• Nonmetal	• Nonmetal
• 2:7 ratio with O	• 2:7 ratio with O	• Does not react with O
• 1:1 ratio with H	• 1:1 ratio with H	• Does not react with H
• Yes (water)	• Yes (water)	• No reaction (water)
• -7°C	• 114°C	• -249°C
• Red brown	• Dark violet	• Colorless

Element G	Element H	Element I
• 85	• 119	• 16
• Solid	• Solid	• Gas
• Metal	• Metal	• Nonmetal
• 2:1 ratio with O	• 1:2 ratio with O	• n/a
• 1:1 ratio with H	• 1:4 ratio with H	• 1:2 ratio with H
• Violent reaction (water)	• No reaction (water)	• No reaction (water)
• 39°C	• 232°C	• -218°C
• Silvery white/soft	• Silvery gray	• Colorless

Element J	Element K	Element L
• 7	• 79	• 35
• Solid	• Solid	• Gas
• Metal	• Nonmetal	• Nonmetal
• 2:1 ratio with O	• 1:3 ratio with O	• 2:7 ratio with O
• 1:1 ratio with H	• 1:2 ratio with H	• 1:1 ratio with H
• Slow reaction (water)	• No reaction (water)	• Yes (water)
• 181°C	• 221°C	• -102°C
• Silvery white/soft	• Gray	• Yellow green

Continued on Page 99 →

Element Property Cards (Page 2)

Element M	Element N	Element O
• 28 • Solid • Metalloid • 1:2 ratio with O • 1:4 ratio with H • No reaction (water) • 1414°C • Dark gray, bluish	• 84 • Gas • Nonmetal • Does not react with O • Does not react with H • No reaction (water) • -157°C • Colorless	• 19 • Gas • Nonmetal • 2:7 ratio with O • 1:1 ratio with H • Yes (water) • -220°C • Pale yellow
Element P	**Element Q**	**Element R**
• 131 • Gas • Nonmetal • Does not react with O • Does not react with H • No reaction (water) • -112°C • Colorless	• 23 • Solid • Metal • 2:1 ratio with O • 1:1 ratio with H • Rapid reaction (water) • 98°C • Silvery white/soft	• 32 • Solid • Nonmetal • 1:3 ratio with O • 1:2 ratio with H • no reaction (water) • 115°C • Yellow
Element S (Blank)	**Element T (Blank)**	**Element U (Blank)**
• • • • • • •	• • • • • • •	• • • • • • •

Instructions: Your instructor has prepared a <u>box filled with various objects</u>. Working in groups, carefully take the objects out of the box and spread them on the desk. You will notice that each of the objects has been labeled with the symbols of the different atoms that make up the object. Some of the objects are composed of two or more different atoms (they are compounds or mixtures), while some are made primarily of just one type of atom (these are elements). Answer the following questions. When you are done, carefully place the objects back in the box.

Questions:
1) Come up with six different properties (for example, blue, liquid, soft, etc.) that you could use to categorize any two or more of the objects. Complete the table below with the name of the objects and the property that they have in common.

Objects	Property	Objects	Property

2) Separate out the objects that are examples of elements. Pick any two of these objects and list at least ten properties for each. If you get stuck, list properties that you can't directly observe, but that you might want to measure if you had the necessary equipment.

Object	Properties

3) Does there seem to be any simple property you can observe that will let you know whether or not something is an element? Explain.

Instructions: Imagine that there are only 18 known elements. The information and properties for these elements have been summarized on the <u>Element Property Cards</u> that you cut out before class. Your job is to make some sense out of all this information by arranging the cards in a two-dimensional array so that there is some organizational principle for both the columns and the rows. You will need to consider what property you will use to arrange the elements in rows and how you will decide to make a column by starting a new row. You may also decide to use the blank cards if you feel you need to hold a space open in your array. Refer to the sample below for an explanation of the information contained on the cards. Do NOT use a periodic table when you are doing this. When you are done, answer the questions.

Sample Card:

Element A
• 12
• Solid
• Nonmetal
• 1:2 ratio with O
• 1:4 ratio with H
• No reaction (water)
• 3550°C
• Black

Information:

Code Letter of Element
• Relative mass (compared to hydrogen)
• State at room temperature (solid, liquid, or gas)
• Metal, nonmetal, or metalloid
• Ratio when combining with oxygen
• Ratio when combining with hydrogen
• Reactivity with water (normal conditions)
• Melting point
• Color

Questions:

1) Below in the space to the left, draw a schematic of your arranged cards (you can just draw boxes with the code letter of the element). Save the space to the right to draw the scheme that the class comes up with if it is different from yours.

2) What organizing principle or property did you use to arrange the elements in rows from left to right?

Continued on Next Page →

3) What organizing principle did you use to arrange the elements in columns from top to bottom? In other words, how did you decide when to make a column by starting a new row?

4) Explain how you used any of the blank cards in your organizational scheme.

5) If you used the blank cards, complete one of the Element Property Cards below for each missing element. This will require you to make a hypothesis concerning each of the properties on the card based on any trends that you can observe.

Element S	Element T	Element U
•	•	•
•	•	•
•	•	•
•	•	•
•	•	•
•	•	•
•	•	•
•	•	•

6) Now, let's look at the periodic table found on the inside front cover of this activity manual. Use the periodic table to match, based on atomic mass, each of the coded elements (A – U) with the name of the actual element.

Code	Element		Code	Element		Code	Element
A			H			O	
B			I			P	
C			J			Q	
D			K			R	
E			L			S	
F			M			T	
G			N			U	

7) How well does you arrangement of Elements A – U compare with the arrangement used in the actual periodic table? Explain.

Instructions: Compare the version of Mendeleev's early periodic table shown below (edited for clarity) to the modern one found on the inside front cover of this activity manual or your textbook. Answer the questions that follow.

Dmitri Mendeleev's 1872 Periodic Table

	1	2	3	4	5	6	7	8
1	H=1							
2	Li=7	Be=9	B=11	C=12	N=14	O=16	F=19	
3	Na=23	Mg=24	Al=27	Si=28	P=31	S=32	Cl=35	
4	K=39	Ca=40	_=44	Ti=48	V=51	Cr=52	Mn=55	Fe=58, Co=59, Ni=59, Cu=63
5		Zn=65	_=68	_=72	As=75	Se=78	Br=80	
6	Rb=85	Sr=87	Y=88	Zr=90	Nb=94	Mo=96	_=100	Ru=104, Rh=104, Pd=106, Ag=108
7		Cd=112	In=113	Sn=118	Sb=122	Te=125	I=127	
8	Cs=133	Ba=137	_	Ce=140				– – – –
9		_	_	_	_	_	_	
10	_	_	_	_	Ta=182	W=184	_	Os=185, Ir=197, Pt=198, Au=198
11		Hg=200	Tl=204	Pb=207	Bi=208	_	_	
12		_		Th=231	_	U=240		

Questions:

1) How many elements were known in 1872 when Mendeleev organized this early version of the periodic table? How many are known today and are represented on the modern periodic table?

   1872:   Present:

2) Mendeleev's version of the periodic table does not have noble gases. Why do you think the noble gases were late in being discovered? Hint: Look at your Element Property Cards from *Part B*.

3) When new elements are added to the periodic table, they have either been discovered or created. Explain the difference.

   Discovered:

   Created:

4) In addition to having identified more elements, the modern periodic table also gives atomic masses with more digits. Why do you think that is the case?

Continued on Next Page →

5) Mendeleev left blanks in his periodic table (indicated by small dashes).
   a) Why, for example, did he put As, the next heaviest known element after Zn, in group 5 instead of in groups 3 or 4?

   b) What did he think the blank spaces suggested?

   c) Was his thinking correct? In other words, what elements did his blanks in columns 3 and 4 (with predicted masses of 68 and 72, respectively) turn out to be?

   d) Are there similar blanks on the modern periodic table between magnesium (Mg) and aluminum (Al)? Explain.

6) Mendeleev used atomic mass to organize his periodic table. Does the modern periodic table also use atomic mass for organizing the elements? If not, what does it use? Hint: Did you notice anything unusual when you tried to organize Elements B and E (*Part B*)? Explain.

Name: _____   Score: _____   points

1) Write an analogy that uses common things to illustrate the relationship between atoms and elements (for example, bricks are to brick walls as atoms are to elements). Briefly explain why your analogy is appropriate and what some of its limitations are.

2) Though you will not need to memorize the whole periodic table, you should learn a few of the more important "vocabulary" terms so that we can communicate effectively. With this in mind, **learn the symbols and names** (spelling counts) of the 20 common elements listed below. To help you start to learn them, complete the following table.

Symbol	Name		Symbol	Name
H			Ca	
He			S	
C			Cl	
N			F	
O			Br	
Al			I	
Li			Fe	
Na			Cu	
K			Sn	
Mg			Pb	

3) Unlike "C" for carbon and "O" for oxygen, some of the elements in the table you just completed have symbols that don't appear to match their English names. For two of the elements, research the origin of their symbol/name and record the source of your information.

Element 1:

Element 2:

Source:

Continued on Back →

4) Represent the data found in the table below in the graphical form of your choice (for example, bar graph, pie chart, etc.). You will need to decide at what point the "percents" are too small to show on your graph and should be grouped as "other elements."
   a) Attach your graph to your Post-Activity Problem Set.
   b) What percent of the currently known elements are essential or thought to be essential to human life?

   c) What do you think is meant by a "trace" percent?

   d) It may seem strange to think that we are 64.6% oxygen. What form do you think most of the oxygen in our bodies takes?

Percent by Mass of Elements in the Human Body

Element	Percent	Function
Oxygen	64.6	
Carbon	18.0	
Hydrogen	10.0	
Nitrogen	3.1	
Calcium	1.9	Found in compounds in bones, teeth, and body fluids.
Phosphorus	1.1	About 85% found in combination with calcium in bones and teeth. Remainder in compounds in body fluids and DNA and RNA in cells.
Potassium	0.36	A major element found in compound contained in cellular fluids. Also involved in transmission of nerve impulses.
Sulfur	0.25	Found in amino acids and proteins of the body.
Chlorine	0.15	Mainly found as dissolved salt contained in extracellular fluids. Also found in gastric juices in the stomach.
Sodium	0.15	Mainly found as dissolved salt in extracellular fluids. Also found in cellular fluids and involved in transmission of nerve impulses.
Magnesium	0.03	Found in compounds contained in bone and body fluids.
Iron	0.004	An important component of blood hemoglobin and muscle myoglobin. Stored in compounds found in the liver, spleen, and bone.
Chromium	Trace	Related to function of insulin in glucose metabolism.
Cobalt	Trace	Required for function of several enzymes, part of vitamin B_{12}.
Copper	Trace	Required for function of some enzymes.
Iodine	Trace	Located in the thyroid gland, needed for hormone thyroxine.
Manganese	Trace	Required for function of several digestive enzymes.
Molybdenum	Trace	Required for function of several enzymes.
Zinc	Trace	Required for function of many enzymes.
Fluorine	Trace	Found in bones and teeth; believed essential, but function unknown.
Selenium	Trace	Essential for liver function.
Silicon	Trace	May be essential in humans.
Tin	Trace	May be essential in humans.

Explorations in Conceptual Chemistry: Activity 3

Unit 4: The Structure of the Atom

In Unit 3, we saw that the periodic table arranges the known elements by placing those with similar properties in the same group. Now, however, we might ask, "*Why* are helium (He), neon (Ne), and argon (Ar) all gases at room temperature," or "*Why* do sodium (Na), potassium (K), and rubidium (Rb) react with water?" Explaining *why* the periodic table works requires knowledge about the internal structure of the atom. Since we can't even directly see the atom, let alone see *inside* the atom, our understanding must again come from interpretation of the results of macroscopic observations. For the activities in Unit 4, we will be relying on the experimental results and observations of others. What we do with the results, however, will be the same. It will be our job to analyze the observations and explain how they lead to a particular conclusion.

In Unit 4, we learn how scientists were able to figure out that the electrons move through the empty space surrounding the small, very dense nucleus. We will also develop a simplified model of the electronic structure of the atom. This model is called the "shell model" and is based on experimental data. The shell model allows us to answer the "*Why*" questions in which we are interested, while avoiding the mathematics and abstract concepts inherent in other, more complicated models of the atom.

Activity 4a: The Nucleus, Isotopes, and Atomic Mass
COVER SHEET

Before Class
- Read the pages in your textbook dealing with "atomic nucleus," "atomic mass," and "isotopes." Your instructor may assign specific pages for you to read.
- Complete the Pre-Activity Problem Set on the next page.

Key Concepts
- Understanding the inner structure of the atom can provide us with insight into the properties of the different elements.
- Since we can't see inside the atom, our understanding of the submicroscopic world must come from interpretation of the results of macroscopic observations.
- The currently accepted model of the atom describes the electrons moving in mostly empty space around a small, dense nucleus.
- Isotopes are atoms of the same element with different numbers of neutrons.
- An element's atomic mass (as reported on the periodic table) is the weighted average of the masses of its isotopes.

Learning Objectives
- You will be able to differentiate between the results and the interpretation of the results.
- You will be able to describe the basic features of Rutherford's gold-foil experiment and his nuclear model of the atom.
- You will be able to explain what an isotope is.
- You will be able to perform calculations related to the atomic mass of an element.

Name: _____   Score: _____ points

1) What does a scientist mean when he or she refers to the **results** (or observations) of an experiment? How about the **interpretation** (or conclusions) of an experiment?

The **Results** of an Experiment Are:

The **Interpretation** of an Experiment Is:

2) With regard to Rutherford's gold-foil experiment, what were the **expected results** and the **actual results** and how did Ernest Rutherford **interpret** those results?

Expected Results:

Actual Results:

Interpretation of Results:

Continued on Back →

3) If the three large circles below represent the size of an atom, which of the following best represents the relative size of the atomic nucleus? Briefly explain your choice.

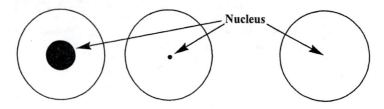

4) Carbon has two stable, naturally occurring isotopes, carbon-12 and carbon-13 (also called $^{12}_{6}C$ and $^{13}_{6}C$, respectively).

a) Regardless of which isotope it is, how many protons does an atom of carbon always have?

b) What do the "12" and "13" tell us about the carbon isotopes?

c) Use your answer to Questions #4a and #4b to determine the number of neutrons in each isotope of carbon.

d) What is the name given to the "number of protons" in an element?

e) What is the name given to the "number of protons + neutrons" in an element?

f) Draw pictures of the carbon-12 and the carbon-13 nuclei. Use solid circles (●) to represent the protons and open circles (O) to represent the neutrons.

Carbon-12 Nucleus   Carbon-13 Nucleus

Instructions: Your instructor will perform a demonstration that can serve as a model for Rutherford's gold-foil experiment.

Observations: Briefly describe the demonstration. Your description should include a labeled drawing of the set-up and written statements of the results and their interpretation.

Questions:

1) How does each part of the demonstration relate to each part of the set-up for Rutherford's experiment?

2) Assuming that the diameter of the atom doesn't change, if the nucleus of the atom were twice as large, would Rutherford have observed more or fewer deflected alpha particles during his gold-foil experiment? Explain.

Instructions: Your instructor has prepared <u>sets of small baggies (lettered A, B, and C)</u>, each filled with two different-colored objects. The baggies represent the three different nuclei for the isotopes of a single element. Without opening the baggies, examine the contents and answer the following questions.

Questions:
Since the baggies represent the same element, they have the same number of protons.
1) Which of the objects in the baggie represents the protons?

2) How many protons does this element have?

3) Using the periodic table, what element has this number of protons?

4) How many neutrons does each of the isotopes (A, B, and C) have?

	Baggie A	Baggie B	Baggie C
# of Neutrons			

5) Based on the identity of the element and the number of protons and neutrons, write the symbol for each isotope.

	Baggie A	Baggie B	Baggie C
Isotope Symbol			

6) Imagine you are asked to make similar models for the four isotopes of sulfur (S-32, S-33, S-34, and S-36). Fill in the table below with the corresponding numbers of protons and neutrons.

	Sulfur-32	Sulfur-33	Sulfur-34	Sulfur-36
# of Protons				
# of Neutrons				
Atomic Number				
Mass Number				

Instructions: In today's activity, we will be using the concept of a "weighted average." As a student, you may be more familiar with weighted averages than you think. Most Grade Point Averages (GPAs) are weighted averages where grades earned for courses worth 4 or 5 units are weighted more heavily than courses worth 2 or 3 units. To help prepare us for understanding what is meant by a weighted average, begin by calculating the average of the ten numbers in each of the sets (#1 – #4) below. When you are done, answer the questions that follow.

	Set #1	Set #2	Set #3	Set #4
	1, 1, 1, 1, 1,	1, 1, 1, 1, 1,	1, 1, 1, 1, 1,	1, 2, 2, 2, 2, 2,
	1, 1, 1, 1, 1	1, 1, 1, 2, 2	2, 2, 2, 2, 2	2, 2, 2, 2
Average				

Questions:

1) How does the average value for each set change as more 2's are added?

2) What fraction of the values in Set #2 are "1"?

3) What fraction of the values in Set #2 are "2"?

4) In cases where there are a lot of values that are the same (here, lots of 1's and 2's), how can your answers to Questions #2 and #3 be used to make a shortcut for calculating the average?

5) Use your shortcut in Question #4 to find the average for a set of 40 numbers (twelve 2's and twenty-eight 3's). Show all your work.

Questions:

1) Imagine that you have a large jar of pennies and you want to know how much it is worth, but you don't have time to sit there and count all the change. Assuming that you have all the tools of a typical chemistry lab at your disposal, propose a method that would allow you to determine the value of the pennies in the jar.

2) It turns out that the method you proposed in Question #1 might not work because not all pennies are identical:
   - Pre-1982 pennies were an alloy composed of 95% copper and 5% zinc. The average pre-1982 penny has a mass of 3.11 grams.
   - Post-1982 pennies consist of a zinc core, coated with a thin layer of copper (overall 97.6% zinc and 2.4% copper), with an average mass of 2.50 grams.

   Knowing this about pennies, how are pennies like isotopes?

3) Examine the contents of a <u>bag of pennies labeled "A."</u> In the bag is a sample of pennies representing the penny's "naturally occurring ratio of isotopic abundance."
   a) How many total pennies are in the bag?

   b) How many pennies are pre-1982?

   c) How many pennies are post-1982?

   d) What fraction of a "naturally occurring" sample of pennies are pre-1982?

   e) What fraction are post-1982?

Continued on Next Page →

4) Using the masses provided in Question #2 and the answers you gave in Questions #3d and #3e, what is the weighted average mass of a penny in the "naturally occurring" sample in Bag A? Show all your work.

5) What would you expect for the mass of a sample of 350 pennies, assuming that they had the same "naturally occurring" ratio of pre- and post-1982 pennies? Show your work.

6) A sample of pennies in a baggie is found to have a mass of 207.67 g. The baggie itself has a mass of 2.84 g. How many pennies are in the sample? How many of the pennies are pre-1982 pennies? Show your work.

7) How is the mass of an average penny (the weighted average) like the atomic mass of an element?

Questions:

1) If there are only two isotopes of K (K-39 with a mass of 39.0 amu and K-40 with a mass of 40.0 amu), which isotope is naturally present in a higher percentage? Explain your answer. Hint: Use the information available on the periodic table.

2) The following questions refer to the element boron, B, which has two naturally occurring isotopes, B-10 (mass of 10.0 amu) and B-11 (mass of 11.0 amu).

   a) List two ways in which an atom of B-10 is similar to an atom of B-11.

   b) List two ways in which an atom of B-10 is different from an atom of B-11.

   c) If B-10 has a 20.0% abundance, what is the percent abundance of the other isotope?

   d) *Without using a calculator*, determine the atomic weight of boron. Clearly show all of the work involved in your calculation.

3) The element chlorine has two naturally occurring isotopes. Knowing that the lighter isotope has a mass of 34.969 amu and an occurrence of 76.0%, answer the following questions.

   a) What is the atomic mass of chlorine according to the periodic table?

   b) What is the percent occurrence of the heavier isotope?

   c) What must be the mass of the heavier isotope? Show all of your work.

Name: _____   Score: _____ points

1) If your pencil is made up of atoms, and according to Rutherford's gold-foil experiment atoms are mostly empty space, why does your pencil feel solid when you pick it up? (Hint: Which part of the atoms in your finger approaches the outermost part of the atoms in your pencil when you go to pick it up?)

2) Which has more atoms: a 1-gram sample consisting of only Cu-63 atoms or a 1-g sample consisting of only Cu-65 atoms? Explain your answer by developing an analogy to this question using Styrofoam balls and metal balls of the same size.

3) A sample of naturally occurring copper is found to consist of two different isotopes. The heavier isotope (Cu-65) has a mass of 64.93 amu and an occurrence of 30.91%. The lighter isotope (Cu-63) has a mass of 62.93 amu. Use this information to calculate the average atomic mass of a copper atom. Where does this information appear on the periodic table if you wanted to check the result of your calculation? Show all your work.

Activity 4b: The Electrons and the Shell Model
COVER SHEET

Before Class

- Read the pages in your textbook dealing with "electrons" and "atomic spectrum." Your instructor may assign specific pages for you to read.
- Complete the Pre-Activity Problem Set on the next page.

Key Concepts

- Insight into the submicroscopic world comes from interpretation of experimental results.
- Electrons are negatively charged particles that move around at very high speeds through the empty space surrounding the small, very dense, positively charged nucleus of an atom.
- The amount of energy it takes to remove an electron from an atom (the ionization energy, IE) can be used to develop a model (the shell model) that shows the arrangement of the electrons in an atom.
- The number of outer, chemically active electrons (the valence electrons) in an atom will dictate, to a large degree, the properties of a given element.

Learning Objectives

- You will be able to explain how the shell model is developed from experimental ionization energies.
- You will be able to apply the shell model to draw a representation of any atom.
- You will be able to use the shell model to predict relative ionization energies of any element in either the same period or the same group of the periodic table.
- You will be able to identify the valence electrons in an atom and use this information to draw Lewis dot symbols.

Name: _____   Score: _____ points

1) What is an atomic spectrum?

2) Why is an atomic spectrum useful?

3) How is the energy of an electron in an atom like a person climbing a ladder?

4) A useful model that can help us conceptualize the structure of the atom is called the *planetary model* of the atom. The *planetary model* of the atom is a reference to our solar system. In this model, what part of the atom is like the sun? What part of the atom is like the planets?

5) What are valence electrons and why are they important?

Background: There are two factors determining how strongly two charged particles interact with each other. These factors are the distance between the particles and the magnitude of the charge on the particles. Charges can be positive (e.g., protons) or negative (e.g., electrons). Particles with like charges repel one another and particles with opposite charges are attracted to one another.

Questions:

1) For two oppositely charged particles, what do you predict is the relationship between how far apart they are and how strongly they are attracted to each other? (In other words, are two oppositely charged particles that are close together going to be held more tightly or less tightly than two oppositely charged particles that are farther apart?) Briefly explain your choice.

2) For two oppositely charged particles, what do you predict is the relationship between the magnitude of their charges and how strongly they are attracted to each other? (In other words, are two oppositely charged particles that have small charges going to be held more tightly or less tightly than two oppositely charged particles that have larger charges?) Briefly explain your choice.

3) Rank the following systems beginning with the system where the electron is held most weakly to the system where the electron is held most strongly. Briefly justify your ranking. Note: 1 pm = 10^{-12} m (so, 500 pm = 0.000000000500 m)

   System A: An electron at a distance of 500 pm from a nucleus with a charge of +2

   System B: An electron at a distance of 500 pm from a nucleus with a charge of +3

   System C: An electron at a distance of 700 pm from a nucleus with a charge of +2

   System D: An electron at a distance of 700 pm from a nucleus with a charge of +1

   System _____ < System _____ < System _____ < System _____
   (Most weakly held)   (Most tightly held)

4) For which of the systems in Question #3 do you think it would take the most energy to remove the electron? Explain.

Instructions: In order to understand the behavior (including the characteristic atomic spectrum) of each element, we need to develop a model for the arrangement of the electrons within the atom. An important tool for developing this model is the experimental data that describes how relatively hard or easy it is to remove an electron from an atom. Use the space below to take notes as your instructor provides you with a brief lecture on photoelectron spectroscopy (PES).

Definition of Ionization Energy (IE):

Summary of How IE is Determined Using PES:

Instructions: Your instructor will provide you with the remaining values for the table below as you develop the shell model for the electronic structure of the atom.

Ionization Energy (IE) for Elements 1 – 20

Symbol	Atomic Number	Ionization Energy (kJ/mol)	Symbol	Atomic Number	Ionization Energy (kJ/mol)
H	1	1310	Na	11	
He	2	2370	Mg	12	
Li	3		Al	13	
Be	4		Si	14	
B	5		P	15	
C	6		S	16	
N	7		Cl	17	
O	8		Ar	18	
F	9		K	19	
Ne	10		Ca	20	

Questions:

1) Why does it make sense that removing an electron from an atom of helium is roughly twice as hard as removing an electron from an atom of hydrogen?

2) Use the space below to draw the models you develop during the class discussion. Remember that the atomic number is the number of protons an atom of that element has. It is also the number of electrons that the neutral atom has.

Hydrogen Atom: Helium Atom:

3) Based on the IE for H and He, write your prediction for the IE for Li in the table above. Use your prediction to draw a model of the Li atom in the space below. When you are done, your instructor will provide you with the remaining IE for the table (including the actual value for Li). Record these values and use this new information to draw a revised model of the Li atom in the space below.

Lithium Atom (predicted):   Lithium Atom (revised):

Instructions: Graph the ionization energies for the first 20 elements (use the data in *Part C*) on a piece of graph paper (found at the end of this manual). Connect the dots on the graph (you are NOT drawing a best-fit line in this case). When you are done, answer the following questions.

Questions:
1) Between which elements was the first drop in IE indicating that the first shell was full and we needed to go on to a second shell? How large was this first decrease in IE?

2) Based on your answer to Question #1, how many electrons did we conclude can fit in the first shell?

3) Between which elements does the next decrease in IE occur, indicating that the second shell is full and we need to go on to a third shell? When answering this question ignore small dips in IE (you are looking for a drop that is as large as the one you noticed in Question #1).

4) Based on your answer to Question #3, how many electrons would you conclude can fit in the second shell?

5) Between which elements does the next decrease in IE occur indicating that the third shell is full and we need to go on to a fourth shell? Again, ignore any small dips in IE.

6) Based on your answer to Question #5, how many electrons would you conclude can fit in the third shell?

7) If you were told (you don't have the values for the IE) that the next large drop in IE occurs between Kr and Rb, how many electrons would you conclude can fit in the fourth shell?

8) Look at the periodic table. Is there any correlation between the number of electrons in each of the first four shells and the number of elements in each of the first four periods? Explain.

9) Fill in the table, summarizing the number of electrons in the seven shells. To fill in the fifth, sixth, and seventh shells, assume that the relationship in Question #8 continues. Remember when determining the number of electrons in the sixth and seventh shells that the lanthanides and actinides belong after Elements 57 and 89, respectively.

Shell	Number of Electrons
1st	
2nd	
3rd	
4th	
5th	
6th	
7th	

Continued on Next Page →

10) Draw the correct number of electrons in each shell of the elements below. Don't forget to write the number of protons in the nucleus next to the "+" charge.

Na, sodium

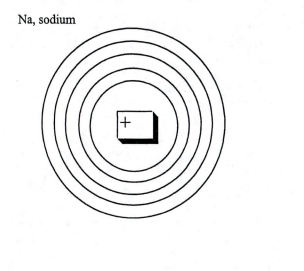

K, potassium

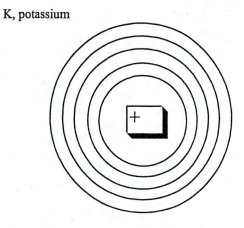

N, nitrogen

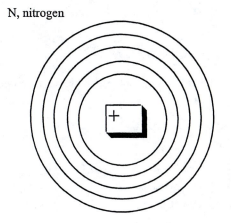

P, phosphorus

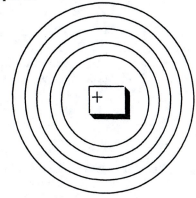

Ne, neon

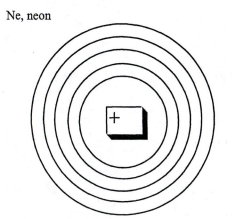

Ar, argon

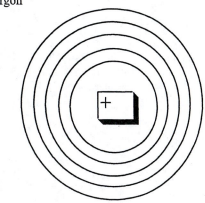

Questions:

1) Look at the shell models you drew in *Part D* and answer the following questions.

   a) Na and K are both in group 1 of the periodic table. How many electrons do they each have in their outermost shell?

   b) Based only on your answer to Question #1a, what *general* pattern might you predict about the number of electrons in the outermost shell (i.e., the **valence electrons**) for any element in group 1?

   c) N and P are both in group 15. How many valence electrons do they both have?

   d) Based only on your answer to Question #1c, what *general* pattern might you predict about the number of valence electrons for any element in group 15?

   e) Ne and Ar are both in group 18. How many valence electrons do they both have?

   f) Based only on your answer to Question #1e, what *general* pattern might you predict about the number of valence electrons for any element in group 18?

   g) Do the answers for Questions #1b, #1d, and #1f suggest a general pattern that can be made about the number of valence electrons for any elements in the same group of the periodic table? Explain.

   h) Does the statement in Question #1g *appear* to hold true for He? Does this statement *appear* to hold true for Br? Explain. (Note: Your instructor will provide you with further information concerning this apparent problem.)

Continued on Next Page →

2) Since the valence electrons are so important, scientists indicate how many an element has by using a shorthand notation called **Lewis dot symbols**. Lewis dot symbols start with the symbol for each element surrounded by a dot for each valence electron. Based on the samples below, draw the Lewis dot symbols for each of the elements in the "Your Turn" section below.

Samples:

➤ Lithium has one valence electron. ➤ It doesn't matter where you draw the electron (top, bottom, left, or right). Li	➤ Calcium has two valence electrons. ➤ Again, it doesn't matter where you draw them, though we don't usually "pair" them until all the sides have one electron. Ca not Ca
➤ Phosphorus has five valence electrons. ➤ Draw one electron on each side before pairing any. P	➤ Neon has eight valence electrons, which is the maximum number. ➤ All the electrons are paired. Ne

Your Turn:

Potassium, has ____ valence electron(s). K	Magnesium has ____ valence electron(s). Mg	Boron has ____ valence electron(s). B
Sulfur has ____ valence electron(s). S	Aluminum has ____ valence electron(s). Al	Fluorine has ____ valence electron(s). F

3) Based solely on their positions on the periodic table, draw the Lewis dot symbols for two more elements that have the same number of valence electrons (and therefore, similar properties) as fluorine, F.

Directions: Your instructor will show you a demonstration in which you will observe the light emitted by various elements when they have been excited by an energy source.

Element	Color of Light Emitted

Questions:

1) Complete the shell models below for hydrogen and helium. Use the extra empty shell to show how one of the valence electrons can be excited to a higher energy shell if it is supplied with energy. Also indicate in your drawing that the same energy is given off in the form of light when the electron falls back down to its original shell.

Hydrogen   Helium

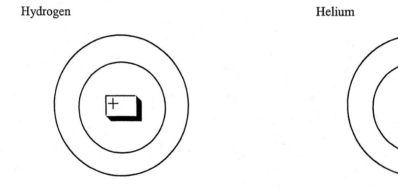

2) How can the shell models above be used to explain the emission of different-colored light by excited atoms of hydrogen and helium?

Name: _____ **Score:** _____ **points**

1) Which factor—distance from the nucleus or charge on the nucleus (i.e., the number of protons)—seems to be the more important factor to consider when determining the ionization energy (IE)? Support your answer with data from the table in *Part C*.

2) Draw the shell model for the following elements:

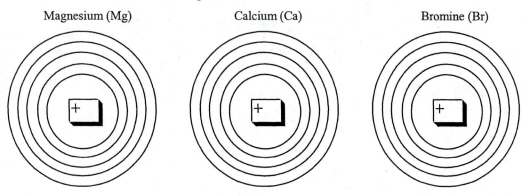

Magnesium (Mg)   Calcium (Ca)   Bromine (Br)

3) Based on your drawings in Question #2, which element would you predict to have a greater ionization energy (harder to remove an electron)—Mg or Ca? Explain.

4) Based on your answer to Question #3, circle the correct choice to complete the following statement:

"Ionization energy increases/decreases as you move from top to bottom in any group of the periodic table."

Continued on Back →

5) Based on your drawings in Question #2, which element would you predict to have a greater ionization energy (harder to remove an electron)—Ca or Br? Explain.

6) Based on your answer to Question #5, circle the correct choice to complete the following statement:

"In general, ionization energy increases/decreases moving from left to right across any period of the periodic table."

7) Mendeleev knew that the odd looking form of the periodic table worked for organizing the properties of the known elements, but he had no idea *why* it worked. For example, he observed that after any similar pair of elements in a given group (for example, after lithium and sodium), that the very next elements (in this case, beryllium and magnesium) would also be similar. Although he believed this pattern was too regular to be a coincidence, he couldn't explain why similar elements keep lining up in groups. How can our understanding of the shell model of the atom be used to explain *why* the periodic table works as a tool for organizing the elements according to their properties?

Unit 5: An Introduction to Ionic, Covalent, and Metallic Bonding

Very little of the matter around us exists as pure elements. Nearly everything is made by combining two or more elements. Even a gold ring has other elements added to make the soft metal stronger. In Unit 5, we will have our first exposure to the various types of compounds that result when elements combine. We will see that the properties of a compound are due in large part to the type (metallic, covalent, or ionic) of bonding present. The type of bonding, in turn, is determined by the type of atoms (metal or nonmetal) that are combining. As with our earlier activities, we will be turning to macroscopic observations (in this case, observations related to electrical conductivity) to gain insight into what is happening in the world of submicroscopic particles.

Activity 5: Conductivity and Models of Chemical Bonding
COVER SHEET

Before Class
- Read the pages in your textbook dealing with "covalent, ionic, and metallic bonding." Your instructor may assign specific pages for you to read.
- Complete the Pre-Activity Problem Set on the next page.
- Bring your textbook to class for this activity.

Key Concepts
- Most of the matter around is a combination of two or more different types of atoms.
- Atoms join together to form compounds with the goal of obtaining a full shell of valence electrons.
- The type of atoms (metals or nonmetals) that make up a compound determine the type of bonding (metallic, ionic, or covalent) found in the compound.
- The type of bonding found in the compound determines the properties of the compound.
- Electrical conductivity requires mobile, charged particles.

Learning Objectives
- You will be able to draw shell models for ions.
- You will be able to determine the typical charge on an ion.
- You will be able to identify the type of bonding present in a compound.
- You will be able to compare and contrast the three types of bonding.
- You will be able to list and explain important properties of the three types of compounds.
- You will be able to support your explanations of the properties of different compounds by drawing submicroscopic representations.

Name: _____   Score: _____ points

1) What is a compound?

2) When and where are subscripts used when writing a chemical formula?

3) Write the chemical formula for ammonia. What does the chemical formula tell us about this compound?

4) Using sodium chloride (NaCl) as an example, explain how the properties of a compound differ from the properties of the elements that make it up.

5) What is an ion?

Continued on Back →

6) Draw the shell model for the following atoms and ions. Also show your work for the calculation of the "net charge" in each case.

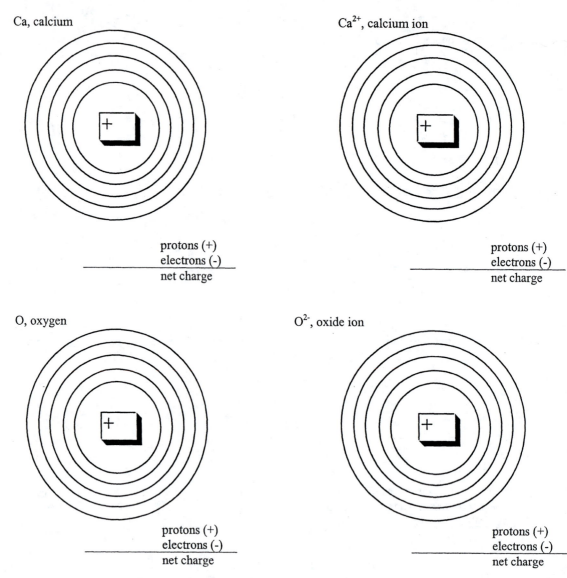

Ca, calcium

protons (+)
electrons (-)
_____
net charge

Ca²⁺, calcium ion

protons (+)
electrons (-)
_____
net charge

O, oxygen

protons (+)
electrons (-)
_____
net charge

O²⁻, oxide ion

protons (+)
electrons (-)
_____
net charge

7) Because the noble gases all have full shells and they are also very stable (they don't react with other elements), we can conclude that there is something advantageous about having a full shell. Notice that when the two atoms in Question #5 became ions, they each ended up with a full outer shell. Why does it make sense that oxygen achieved a full shell by gaining electrons, while calcium did it by losing electrons?

Part A: Types of Bonding (Student Presentations)

Instructions: Working in pairs or small groups, your instructor will assign you one of the three types of bonding (metallic, covalent, or ionic) to research. Your group will have 30 minutes to prepare a 10-minute presentation (see the five bullets below and on the next page) on your assigned type of bonding. After the presentations are prepared, your instructor will call on one group for each type of bonding to give their 10-minute presentation to the rest of the class. Be prepared to take notes and ask questions concerning the presentations you observe on the other two types of bonding.

Group 1: Metallic bonding

- Type of Element Required (Metals, Nonmetals, or a Combination):

- Examples (Chemical Formulas):

- Description of the Nature of the Bond:

- Submicroscopic-Scale Representation (I.E., a Drawing) of a Material with This Type of Bonding:

- Macroscopic Properties of Materials with This Type of Bonding:

Continued on Back →

Group 2: Ionic Bonding

- Type of Element Required (Metals, Nonmetals, or a Combination):

- Examples (Chemical Formulas):

- Description of the Nature of the Bond:

- Submicroscopic-Scale Representation (I.E., a Drawing) of a Material with This Type of Bonding:

- Macroscopic Properties of Materials with This Type of Bonding:

Continued on Next Page →

Group 3: Covalent Bonding

- Type of Element Required (Metals, Nonmetals, or a Combination):

- Examples (Chemical Formulas):

- Description of the Nature of the Bond:

- Submicroscopic-Scale Representation (I.E., a Drawing) of a Material with This Type of Bonding:

- Macroscopic Properties of Materials with This Type of Bonding:

Part B: Does It Conduct Electricity?

Instructions #1: Your instructor will show you a <u>conductivity probe</u>, and the class will discuss how it might be used to determine the electrical conductivity of a sample of matter. Based on the discussion, answer the questions below.

Observations:
Sketch a labeled representation of the conductivity probe you will be using.

Questions:
1) What two things are required for something to conduct electricity?

2) What are a few ways that the "open" circuit on the conductivity probe could be "closed"?

3) Discuss how we can use macroscopic observations from the conductivity probe to gain insight into the submicroscopic level of the samples of matter we will be testing.

Continued on Next Page →

Instructions #2: Prepare the <u>samples below</u> according to the **directions found on page 149**. Test the electrical conductivity of each sample using the <u>conductivity meter</u>. Electrical conductivity is determined by observing the light bulb on the conductivity meter. The brighter the light, the greater the electrical conductivity of the sample. If no light is observed, the solution does not conduct electricity. Be sure to rinse the electrode with distilled water and dry it between testing different samples to avoid cross-contamination. When you are finished, dispose of the samples according to your instructor's directions. Clean all glassware and the electrodes with distilled water.

Set	Name of Sample	Formula	Type of Bonding	Does It Conduct?
#1	Copper wire	$Cu(s)$		
	Aluminum foil	$Al(s)$		
#2	Pure water (distilled)	$H_2O(l)$		
	Tap water	$H_2O(l) + ???$	???	
#3	Paraffin (candle wax)	$C_{30}H_{62}(s)$		
	Plastic	Made of C and H		
	Sucrose (table sugar)	$C_{12}H_{22}O_{11}(s)$		
#4	Sodium chloride (table salt)	$NaCl(s)$		
	Sodium chloride dissolved in pure water	$NaCl(aq)$		
#5	Pure ethanol	$C_2H_5OH(l)$		
	Rubbing alcohol (alcohol dissolved in pure water)	$C_3H_7OH(aq)$		
#6	Magnesium sulfate (Epsom salt)	$MgSO_4(s)$		
	Magnesium sulfate dissolved in pure water	$MgSO_4(aq)$		

Continued on Page 149 →

Directions for Sample Preparation and Testing

Set	Sample	Directions
#1	Copper wire	Without touching the electrodes to each other, touch the electrodes to the copper wire to test the conductivity.
	Aluminum foil	Without touching the electrodes to each other, touch the electrodes to the aluminum foil.
#2	Pure, distilled water	Fill an empty well on the well plate with distilled water. Test the water for conductivity.
	Tap water	Fill an empty well on the well plate with tap water. Test the water for conductivity.
#3	Paraffin (candle wax)	Without touching the electrodes to each other, gently insert the electrodes into the wax to determine the conductivity.
	Plastic	Without touching the electrodes to each other, touch the electrodes to the plastic.
	Sucrose (table sugar)	Place a small amount of the solid (just the end of a spatula) in a clean, dry, empty well on the well plate. Without touching the electrodes to each other, touch the electrodes to the solid to determine the conductivity
#4	Sodium chloride (table salt)	Place a small amount of the solid (just the end of a spatula) in a clean, dry, empty well on the well plate. Without touching the electrodes to each other, touch the electrodes to the solid to determine the conductivity.
	Sodium chloride in pure water	Add enough distilled water to the solid so that the well is full. Stir the solid to facilitate dissolving. Submerge the electrodes in the water to test the conductivity.
#5	Pure ethanol	Fill an empty well on the well plate with ethanol. Test the ethanol for conductivity.
	Rubbing alcohol (alcohol in pure water)	Fill an empty well on the well plate with rubbing alcohol. Test the rubbing alcohol for conductivity.
#6	Magnesium sulfate (Epsom salt)	Place a small amount of the solid (just the end of a spatula) in a clean, dry, empty well on the well plate. Without touching the electrodes to each other, touch the electrodes to the solid to determine the conductivity.
	Magnesium sulfate in pure water	Add enough distilled water to the solid so that the well is full. Stir the solid to facilitate dissolving. Submerge the electrodes in the water to test the conductivity.

Instructions: Summarize the results from *Part B* by completing the table below. Circle "Yes" or "No" for each possible combination, depending on whether or not electrical conductivity was observed. In each box, also draw a picture that explains your observations by showing the presence (or absence) of mobile, charged particles.

	Do materials with **ionic** bonds conduct electricity…	Do materials with **metallic** bonds conduct electricity…	Do materials with **covalent** bonds conduct electricity…
when in the **solid** state?	Yes No	Yes No	Yes No
when in the **liquid** state?	Yes No	Yes No	Yes No
when **dissolved in water** (aq)?	Yes No	No drawing is required here. Metals do not dissolve in water unless they undergo a chemical change.	Yes No

Name: Score: points

1) In Activity 2c, you saw that an aluminum rod (Al) conducted heat better than a glass rod (SiO_2). Which do you think would be better at conducting electricity? Explain your answer in terms of the presence (or lack) of mobile, charged particles.

2) What is the formula for baking soda? Would you expect baking soda dissolved in water to conduct electricity? Explain your answer in terms of the presence (or lack) of mobile, charged particles.

3) Based on what you saw in this activity, why do you think we use plastic to coat electrical wires in our homes? Explain your answer in terms of the presence (or lack) of mobile, charged particles.

Continued on Back →

4) Based on our conductivity results, what type of compound (ionic or covalent) must have been dissolved in pure water to give tap water? Explain your answer in terms of the presence (or lack) of mobile, charged particles.

5) After graduating college with a degree in chemistry, you get a job working for a crime lab. For your first robbery case, you have been given a white powder found on footprints left at the scene of the crime. You are told that there are two suspects in the case, one is the pizza delivery man who works with pizza flour (made of C, H, O, and N) and the other is the gardener who works with various fertilizers (made with Na, N, and O).
   a) Describe a simple test based on what you have learned in this activity that could be used to help you identify the white powder and the thief.

   b) What results would you expect for your test if the pizza delivery man is the thief? Explain your answer in terms of the presence (or lack) of mobile, charged particles.

   c) What results would you expect for your test if the gardener is the thief? Explain your answer in terms of the presence (or lack) of mobile, charged particles.

Unit 6: Exploring Covalent Compounds (Molecules)

Many molecules serve important roles as drugs, personal and housekeeping products, and fuels. Knowing the three-dimensional shape of molecules is crucial to understanding a wide variety of applications, ranging from how various proteins work in our body to why water is liquid at room temperature, while carbon dioxide is a gas. Therefore, one of our goals in Unit 6 is to develop a model that will allow us to predict the three-dimensional shape of simple molecules. Once we know the three-dimensional shape of a molecule, we can better understand many of its properties, such as boiling point.

In Unit 6, we'll also spend some time examining the properties of water. Although you've most likely seen water every day of your life (drinking, bathing, swimming, etc.), Activity 6d may be the first time you've ever *really* looked at this most amazing of compounds.

We close Unit 6 with an activity that will introduce you to the chemistry behind the plastics and polymers that have become ubiquitous in our society.

Activity 6a: Molecules and Lewis Dot Structures
COVER SHEET

Before Class

- Read the pages in your textbook dealing with "covalent compounds" and "Lewis dot structures." Your instructor may assign specific pages for you to read.
- Complete the Pre-Activity Problem Set on the next page.
- Cut out the four pages of puzzle pieces found immediately after the Pre-Activity Problem Set. Put them in an envelope for safekeeping and bring them to the lab.

Key Concepts

- Lewis dot structures are a form of symbolic representation that is often successful in illustrating the bonding within covalent (molecular) compounds.
- Covalent compounds (molecules) result when nonmetal elements share electrons in order to achieve a full valence shell.

Learning Objectives

- You will be able to determine the basic rules guiding the drawing of Lewis dot structures.
- You will be able to draw Lewis dot structures for various common molecules and polyatomic ions.

Name: _____   **Score:** _____   **points**

1) What is a covalent bond? How is a covalent bond like peanut butter between two slices of bread?

2) What is a molecule?

3) Complete the Lewis dot symbols for the following elements:

a)	b)	c)	d)
C	O	N	F

4) Indicate which electrons are shared as bonds and which electrons are the nonbonding lone pair(s) in the following drawings:

a)	b)
H : F :	H : O : H

5) Distinguish between the single, double, and triple bonds in the following molecules:

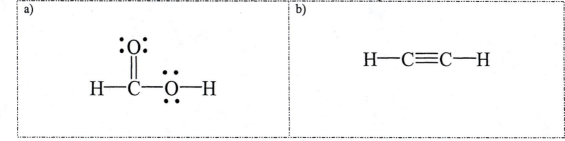

a)	b)
	H—C≡C—H

6) Look at the puzzle pieces on the next four pages (don't forget to cut them out before class). What factor determines how many dots are on each puzzle piece?

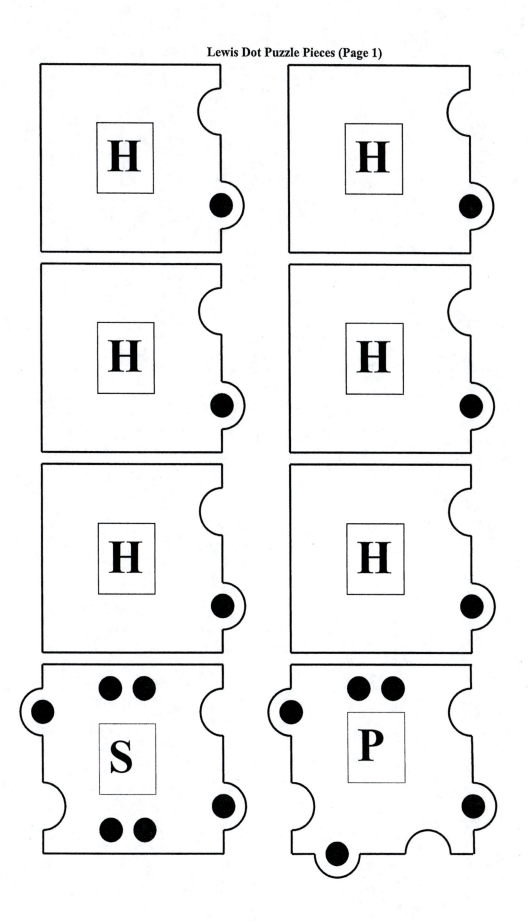

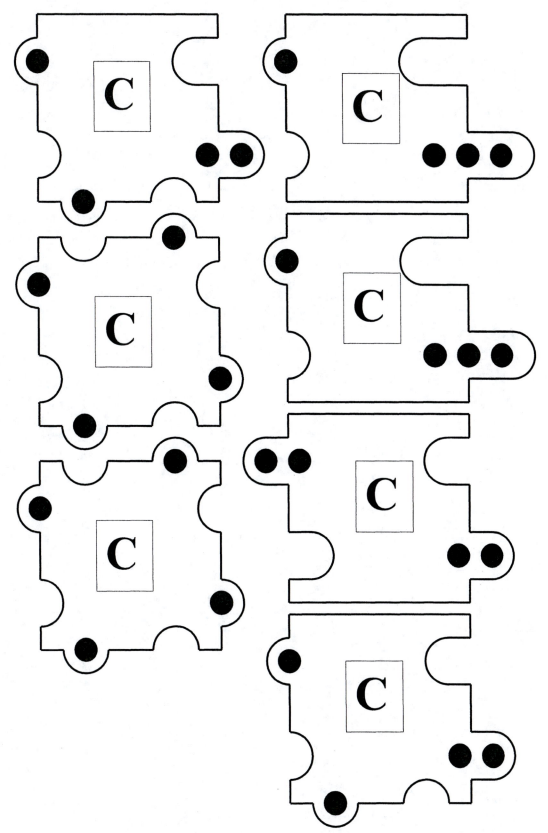

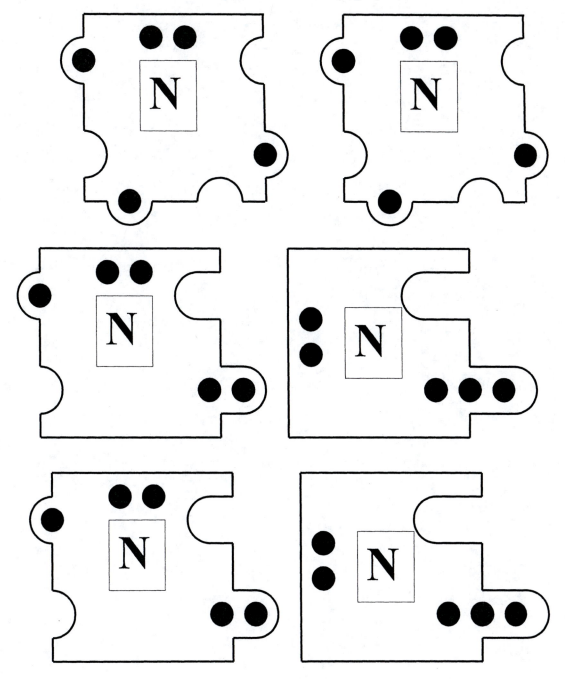

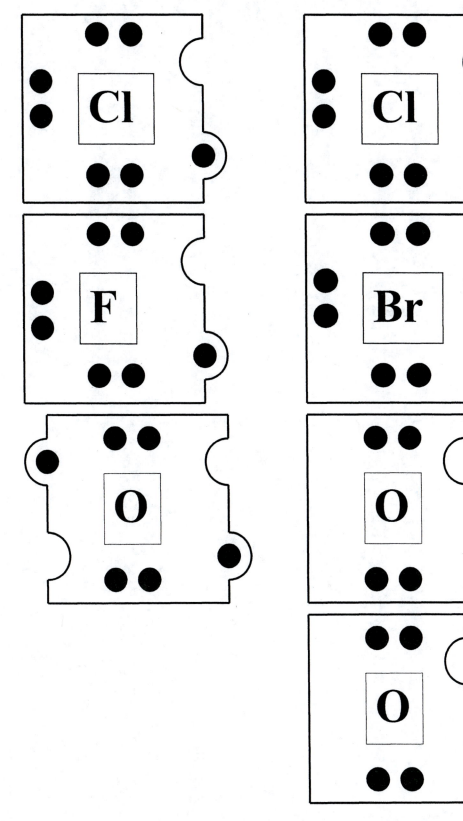

Part A: Lewis Dot Structures (for Molecules with Single Bonds)

Instructions #1: Working in pairs, use the puzzle pieces you cut out to build, one at a time, the Lewis dot structures for the covalent molecules below. In each of these compounds only one pair of electrons is shared by any two atoms (this is called a *single bond*). The remaining electrons are called *nonbonding lone pairs* or *unpaired electrons*.

Questions: From your completed puzzle, draw what each molecule looks like by replacing each pair of shared electrons with a line (bond). Use the HF example below as a guide.

Example: HF   becomes:

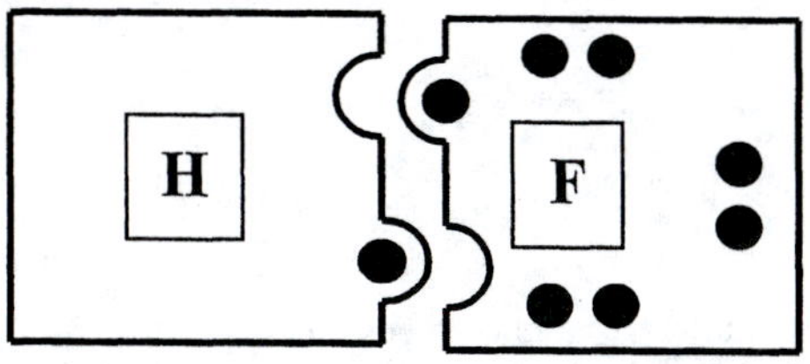

$$H-\overset{\displaystyle \cdot\cdot}{\underset{\displaystyle \cdot\cdot}{F}}\!\cdot\cdot$$

NH₃	H₃CCH₃ (usually written C₂H₆)
CH₂ClBr	H₂S
CH₄	H₂O
PH₃	H₂NNH₂ (usually written N₂H₄)

NH₃ → NH_3

Continued on Back →

Explorations in Conceptual Chemistry: Activity 6a

167

Instructions #2: When the class is done with *Part A*, your instructor will have students draw each of the final molecules on the board. You will use these sample molecules to determine some of the rules governing the formation of covalent compounds.

Questions:

1) How many bonds does C typically form?

2) Does the answer to Question #1 make sense based on the number of valence electrons that C needs to share to obtain a full valence shell? Explain.

3) How many bonds does N typically form?

4) Does the answer to Question #3 make sense based on the number of valence electrons that N needs to share to obtain a full valence shell? Explain.

5) How many bonds does O typically form?

6) Does the answer to Question #5 make sense based on the number of valence electrons that O needs to share to obtain a full valence shell? Explain.

7) How many bonds do H, F, Cl, and Br typically form?

8) Does the answer to Question #7 make sense based on the number of valence electrons that H, F, Cl, and Br need to share to obtain full valence shells? Explain.

9) Based on your answer to Question #7, explain why H, F, Cl, and Br are most likely to be end (terminal) atoms rather than central (bridging) atoms in a molecule.

Instructions #1: Working in pairs, use the puzzle pieces you cut out to build, one at a time, the Lewis dot structures for the covalent molecules below. In each of these compounds, two pairs (a "double bond") or even three pairs (a "triple bond") of electrons may be shared between any two atoms.

Questions: From your completed puzzle, draw what each molecule looks like by replacing each pair of shared electrons with a line (bond). If two atoms share four electrons, then draw two lines (double bond). Use the O_2 example as a guide.

Example: O_2 becomes:

HClCCHCl (usually written $C_2H_2Cl_2$)	N_2
HCN	H_2CCH_2 (usually written C_2H_4)
HNNH (usually written N_2H_2)	HCOOH

Continued on Back →

Instructions #2: When the class is done with *Part B*, your instructor will have students draw each of the final molecules on the board. You will use these sample molecules to determine some of the rules governing the formation of covalent compounds.

Questions:
1) If a "double bond" counts as two bonds and a "triple bond" counts as three bonds, do these compounds all follow the rules you determined from the molecules in *Part A*? Explain.

2) In summary, what is the typical number of bonds that each of the following elements ends up with in a covalent compound?
   a) Nonmetal elements in group 14 (C and Si)

   b) Nonmetal elements in group 15 (N, P, and As)

   c) Nonmetal elements in group 16 (O, S, Se, and Te)

   d) Hydrogen and elements in group 17 (H and F, Cl, Br, and I)

Instructions: Your instructor will show you how to use the Lewis Dot Structure Template. Though you are not required to use this method, it works very well for the molecular compounds you will be dealing with in this class. Once you are proficient at drawing Lewis dot structures, there is no need to draw the "Skeleton," "Skeleton + Shared electrons," etc. in separate steps.

Formula: **CH₂O**

Calculate Number of Shared Electrons:

$$N \quad - \quad A \quad = \quad S$$

$$- \quad\quad =$$

What does "N" stand for?

What does "A" stand for?

What does "S" stand for?

Some Useful Rules for Drawing Skeletons:	Skeleton + Shared Electrons:
• **Symmetrical if possible** • **H and halogens are terminal.** • **Element with fewest valence electrons in center** • **No O—O bonds (except in O₂ and O₃)** Skeleton:	
Skeleton + Available Electrons (I.E., All Shared and Nonbonding Lone Pairs):	Final Structure: **Check Your Work:** • **Total # of electrons = A** • **# of shared electrons = S** • **Atoms have eight electrons (except H has two)**

Continued on Back →

Questions: Draw the Lewis dot structure for the following molecule on your own. When you are done, compare your answer to your partner's answer.

Formula: **CH₃CN**

Calculate Shared Electrons:

N   -   A   =   S

_____|

-                 =

Skeleton:	Skeleton + Shared Electrons:
Skeleton + Available Electrons:	**Final Structure:**

Instructions: Sometimes molecules gain one or more electrons from a metal atom, resulting in a molecule with a negative charge (called a polyatomic ion). Your instructor will now show you how to use the template to draw the Lewis dot structure of polyatomic ions. Don't worry if the polyatomic ions don't follow the rules we developed about how many bonds each element typically forms, since the additional electron(s) can change the rules.

Formula: NO_3^-

Calculate Number of Shared Electrons:

	N	-	A	=	S
		-		=	

How do we account for the negative charge on the polyatomic ion?

Skeleton:

Skeleton + Shared Electrons:

Skeleton + Available Electrons:

Final Structure:

Continued on Back →

Questions: Try the following polyatomic ion on your own. When you are done, compare your answer to your partner's answer.

Formula: CO_3^{2-}

Shared Electrons:

$$N \quad - \quad A \quad = \quad S$$

_____

$-$

$=$

Skeleton:	Skeleton + Shared Electrons:
Skeleton + Available Electrons:	Final Structure:

Name: _____ Score: _____ **points**

1) Look back at the molecules you built in *Part A* of this activity and answer the following:
   a) Explain why carbon, nitrogen, oxygen, and fluorine form covalent bonds with hydrogen to give CH_4, NH_3, H_2O, and HF, respectively. (Hint: Focus on the number of hydrogen atoms needed in each molecule).

   b) Why does it make sense that both nitrogen and phosphorus form compounds with hydrogen in the same 1:3 ratio (i.e., NH_3 and PH_3, respectively)? Based on your answer, what would you predict for the formula for the molecule formed when arsenic (As) reacts with hydrogen?

2) Use the template (or the method of your choice) to draw Lewis dot structures for the following molecules and polyatomic ions. If you use the copies of the blank templates located on pages 177 – 182, staple them to the back of this problem set.
   a) Formula: **CH_3OH**

Continued on Back →

b) Formula: CO_2

c) Formula: CH_3COO^-

d) Formula: C_2Cl_2

e) Formula: NH_4^+

For additional practice, redraw the Lewis dot structures for the molecules in *Parts A* and *B* of this activity using the template method instead of the puzzle pieces.

Lewis Dot Structure Template

Formula:

Shared Electrons:

	N - A	= S
	-	=

Skeleton:	Skeleton + Shared Electrons:
Skeleton + Available Electrons:	Final Structure:
Notes:	

Continued on Back →

Lewis Dot Structure Template

Formula:

Shared Electrons:

N - A = S

- =

Skeleton:

Skeleton + Shared Electrons:

Skeleton + Available Electrons:

Final Structure:

Notes:

Continued on Next Page →

Lewis Dot Structure Template

Formula:

Shared Electrons:

N   -   A   =   S

-   =

Skeleton:	Skeleton + Shared Electrons:
Skeleton + Available Electrons:	**Final Structure:**

Notes:

Continued on Back →

Lewis Dot Structure Template

Formula:

Shared Electrons:

N - A = S

- =

Skeleton:

Skeleton + Shared Electrons:

Skeleton + Available Electrons:

Final Structure:

Notes:

Continued on Next Page →

Lewis Dot Structure Template

Formula:

Shared Electrons:

$$N \quad - \quad A \quad = \quad S$$

$$- \qquad =$$

Skeleton:

Skeleton + Shared Electrons:

Skeleton + Available Electrons:

Final Structure:

Notes:

Continued on Back →

Lewis Dot Structure Template

Formula:

Shared Electrons:

$$N \quad - \quad A \quad = \quad S$$

$$- \qquad =$$

Skeleton:	Skeleton + Shared Electrons:
Skeleton + Available Electrons:	**Final Structure:**
Notes:	

Activity 6b: VSEPR Theory and Molecular Shape
COVER SHEET

Before Class

- Read the pages in your textbook dealing with "valence shell electron pair repulsion theory." Your instructor may assign specific pages for you to read.
- Complete the Pre-Activity Problem Set on the next page.

Key Concepts

- The three-dimensional shape of a molecule is largely determined by the number of electron regions (substituents) on the central atom.
- The relationship between molecular shape and substituents is summarized by VSEPR theory.

Learning Objectives

- You will be able to predict the three-dimensional shape (including bond angles) of various common molecules and polyatomic ions using VSEPR theory.
- You will be able to draw two-dimensional representations of molecules and polyatomic ions using conventions that indicate the three-dimensional nature of the compound.

Name: _____   **Score:** _____ **points**

1) What does the acronym VSEPR stand for? Explain what each part of the name means with regard to the theory it describes.

2) According to VSEPR theory, what is a substituent?

3) How many substituents does each central atom have in the following molecules? Draw a circle around each of the substituents.

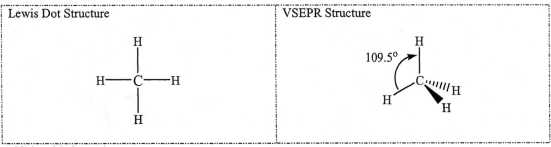

4) Both structures below are symbolic representations for the molecular compound methane, CH_4. Both structures indicate that the carbon is the central atom in the molecule and that there are four hydrogen atoms sharing electrons (covalently bonded) with the carbon. What additional information does the VSEPR structure to the right attempt to convey about the molecule? What is being indicated by the solid arrow and by the dashed arrow in the VSEPR structure?

Lewis Dot Structure	VSEPR Structure
H | H—C—H | H	109.5° → H C''''''H H H

Additional Information:

Solid Arrow:

Dashed Arrow:

Draw the final Lewis dot structures for the following compounds in the boxes provided:

HCN	CO_3^{2-}	CF_4
Show Work:	Show Work:	Show Work:
Final Lewis Dot Structure:	Final Lewis Dot Structure:	Final Lewis Dot Structure:

Circle the central atom in each of the above compounds.
How many substituents (in this case, atoms) are bonded to the central atom in each compound?

# of substituents on
central atom in HCN

# of substituents on
central atom in CO_3^{2-}

# of substituents on
central atom in CF_4

Continued on Back →

To determine the three-dimensional shape of these compounds, VSEPR theory predicts that the substituents (here, the terminal atoms) will try to get as far away from each other as possible. To determine exactly how far apart they can get, take a small ball of clay or a gumdrop (your "central atom") and place two, three, or four toothpicks (your "substituents") in the clay ball as far apart as possible. Draw each clay/toothpick combination in the corresponding box below. Label the angles between the toothpicks. Before going on, have your instructor make sure your drawings are correct. Your instructor may also demonstrate a model using balloons to help show how atoms try to get as far apart from each other as possible.

Drawing of Clay Ball with Two Toothpicks and Labeled Angle:	Drawing of Clay Ball with Three Toothpicks and Labeled Angle:	Drawing of Clay Ball with Four Toothpicks and Labeled Angle:

Now that you know what angle to expect when two, three, or four substituents are bonded to a central atom, go back and draw the three-dimensional VSEPR shapes (with labeled bond angles) of the compounds on the previous page. **Your three-dimensional VSEPR drawing should look just like a complete Lewis dot structure (include all lone pairs as well as single, double, and triple bonds), but in addition, you must now show the correct three-dimensional perspective and have labeled bond angles. To correctly draw four substituents, refer back to the drawing for CH$_4$ on the Pre-Activity Problem Set.**

HCN	CO$_3^{2-}$	CF$_4$

Optional:
If you have molecular model sets available, build each of the above compounds using the model sets. Compare each model with your drawing to see if they are in agreement.

Part B: Shapes of Molecules with One Central Atom (with Lone Pairs on the Central Atom)

According to VSEPR theory, all valence electrons are treated the same, regardless of whether they are used to bond to another atom or are lone pairs. The following compounds all have lone pairs on the central atom.

Lewis Dot Structure	3-D VSEPR Drawing with Labeled Bond Angles
NO_2^-	
H_2O	
NH_3	

Optional:
After drawing each three-dimensional VSEPR shape, you and your lab partner should again build each compound using the molecular model sets. Compare each model with your drawing to see if they are in agreement. Bend two/three of the long connectors (or springs) to make double/triple bonds.

Now that you are an expert at drawing three-dimensional structures of simple molecules, try these more complicated ones. Note that the bond angle around each central atom is determined by the number of substituents on that central atom. Be sure to label the bond angles for each of the central atoms in your final VSEPR drawing.

Lewis Dot Structure	3-D VSEPR Drawing with Labeled Bond Angles
CHCH	
CH_2CH_2	
CH_3CN	

Optional:
After drawing each three-dimensional VSEPR shape, you can again build each compound using the molecular model sets. Compare each model with your drawing to see if they are in agreement.

Post-Activity Problem Set: Activity 6b

Name: _____   Score: _____ **points**

1) For the following molecules and polyatomic ions, draw the Lewis dot structure and the three-dimensional VSEPR structure. **Don't forget to label your bond angles.**

CH_2O
Lewis Dot Structure
3-D VSEPR Structure (with Labeled Bond Angles)

CH_3CH_3
Lewis Dot Structure
3-D VSEPR Structure (with Labeled Bond Angles)

NH_4^+
Lewis Dot Structure
3-D VSEPR Structure (with Labeled Bond Angles)

SO_4^{2-}
Lewis Dot Structure
3-D VSEPR Structure (with Labeled Bond Angles)

Activity 6c: Polarity, Intermolecular Forces, and Boiling Point
COVER SHEET

Before Class
- Read the pages in your textbook dealing with "electronegativity," "polar bonds," and "intermolecular forces." Your instructor may assign specific pages for you to read.
- Complete the Pre-Activity Problem Set on the next page.

Key Concepts
- Polar bonds result when nonmetals form a covalent bond with an unequal sharing of electrons.
- Polar bonds can cancel out to result in a nonpolar molecule.
- Intermolecular forces (IMF) attract molecules to each other in the liquid and solid states. In the gaseous state, the molecules have sufficient kinetic energy to overcome their intermolecular forces.
- Polar molecules are attracted to other polar molecules by permanent dipoles.
- Nonpolar molecules are attracted to other nonpolar molecules by induced (temporary) dipoles.
- In general, given two molecules with similar numbers of electrons, the polar one has the higher boiling point.
- In general, given two molecules that are both polar or both nonpolar, the one with more electrons has the higher boiling point.

Learning Objectives
- You will be able to determine if a given bond is polar.
- You will be able to determine if a given molecule is polar.
- You will be able to determine the nature of the intermolecular forces holding molecules together in the liquid and solid states.
- You will be able to discuss what is happening to the electrons in the continuum of bonding from nonpolar covalent to ionic.
- You will be able to explain trends in boiling points.
- You will be able to explain how soaps and detergents work.

Name: _____   Score: _____ points

1) What is meant by the term *electronegativity*?

2) What is the most electronegative element on the periodic table? What is it about this particular element that makes it so electronegative?

3) The H and F atoms in an HF molecule are held together by the mutual attraction of each atom's nuclei for their shared electrons. The same is true in an F_2 molecule where two F atoms share their bonding electrons. There is a difference between these two molecules, however, in terms of *how* their electrons are being shared. Explain this difference. As part of your explanation, discuss the relative electronegativity of the atoms involved in each of the two bonds.

4) The drawing below depicts two HF molecules and the intermolecular forces that hold them together in the liquid and solid phases. Label both the intermolecular force and any covalent bonds that are shown in the drawing. Also label the molecules using either the dipole ($\delta+$ and $\delta-$) or crossed-arrow notation. Briefly explain how you made your choices.

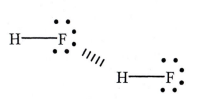

Continued on Back →

5) Imagine that you have two samples, Compound A with a boiling point of -70°C and Compound B with a boiling point -110°C. Starting at -250°C, you begin to slowly heat the samples and the temperature rises.
    a) Which will begin to boil first? Explain.

    b) What do the boiling points allow you to conclude about the strength of the intermolecular forces (the attractions between the molecules) in Compound A versus the attractions between the molecules in Compound B? Explain your answers.

Instructions: Use a periodic table to complete the following table indicating the total number of electrons in the following noble gases.

Element	B.P. (°C)	Total # of Electrons in Atom
He	−269	
Ne	−246	
Ar	−185	
Kr	-153	

Questions:

1) What is the general relationship between the boiling points and the total number of electrons in the noble gases?

2) Based only on boiling points, which of the noble gases has the greatest amount of stickiness (the strongest intermolecular forces) between its atoms? Explain how you made your decision.

Again, complete the following table.

Compound	B.P. (°C)	State at Room Temperature	Total # of Electrons in Molecule
F_2	− 188		
Cl_2	− 34		
Br_2	+ 59		
I_2	+ 184 (sublimes)		

3) Do the boiling points for the halogen molecules follow the same trend we saw in the noble gases? Briefly explain.

4) The intermolecular forces (IMF) between individual noble gas atoms or nonpolar molecules (like the halogens) is called an *induced dipole* (also called "temporary dipoles," "van der Waals forces," or "London dispersion forces"). What is the nature of the *induced dipole* and how does it act to attract neighboring nonpolar molecules?

5) How do *induced dipoles* explain the above trend in noble gas and halogen boiling points?

Instructions: Complete the following table.

Compound	B.P. (°C)	Total # of Electrons
Br_2	+59	
ICl	+97	

Questions:

1) Does the relationship from *Part A* concerning the total number of electrons in a substance and its boiling point hold true with Br_2 and ICl? Explain.

2) The type of intermolecular force (IMF) between polar molecules is called a *permanent dipole*. Use Lewis dot structures for Br_2 and for ICl to show if each molecule is polar or nonpolar. Label any polar molecules using either the dipole ($\delta+$ and $\delta-$) or crossed-arrow notation. Label which molecule would exhibit an *induced dipole* and which would have a *permanent dipole*?

3) Describe the nature of the IMF (induced dipoles) holding Br_2 molecules together in the liquid state. Include a labeled drawing as part of your explanation.

4) Describe the nature of the IMF (permanent dipoles) holding ICl molecules together in the liquid state. Include a labeled drawing as part of your explanation.

Continued on Next Page →

5) Given molecules with similar numbers of electrons, which of these two types of IMFs (*induced dipole* or *permanent dipole*) is stronger? Defend your answer using boiling point data for Br_2 and ICl.

6) In the gas phase, the ICl molecules have enough kinetic energy to overcome their **intermolecular** forces. Are the ICl molecules also moving fast enough to overcome their **intramolecular** bonds (i.e., the covalent bonds) holding the molecules together? Explain.

7) In the left-hand box below, draw a submicroscopic representation of ICl molecules in the liquid phase (remember back to how we draw solids, liquids, and gases). Use dashed lines between the molecules to indicate the intermolecular forces (i.e., the attractive of the $\delta+$ of one ICl molecule for the $\delta-$ of a neighboring ICl molecule). In the box to the right, draw a representation of what the sample would look like if the ICl is heated until it becomes a gas. Make sure your drawing agrees with your answer to Question #6.

+ heat $\Rightarrow$

CAUTION: Hexane is volatile and flammable.

In *Part B,* we saw that we must take into account not only the total number of electrons in the molecule, but also whether the bonds in the molecule are polar or nonpolar. In *Part C,* we explore yet another complication: Though both water (H_2O) and methane (CH_4) are made up of polar bonds (H—O in the case of water and H—C in the case of methane) and have similar numbers of electrons, water has a much higher boiling point. To help us figure out what is going on, consider the following demonstration.

Compound	Total # of Electrons	B.P. (°C)
H_2O	10	+ 100
CH_4	10	- 161

Demonstration: Since CH_4 is a gas, we will use hexane (C_6H_{14}) to model its behavior. Your instructor will rub a balloon on a wool sweater or a volunteer's hair to develop a static charge on the balloon. Your instructor will then hold the balloon alternately near a stream of hexane and then near a stream of water.

Observations for Stream of Hexane and Balloon:

Observations for Stream of Water and Balloon:

Question:
1) Though water and hexane both have polar bonds, only one of them is a polar molecule and the other is a nonpolar molecule. Which molecule do you think is nonpolar, having relatively weak *induced dipoles*? Which of the molecules is polar, having relatively strong *permanent dipoles*? Support your answer using both boiling point data and the results of how they reacted to a charged balloon.

Continued on Next Page →

Instructions: To help us understand how sometimes polar bonds can cancel each other out to give a nonpolar molecule, your instructor will have students perform a series of demonstrations by pulling on lengths of colored yarn tied to a central ball.

Sketch Observations for Two Students (Two Substituents):

Two Identical Substituents:	Two Different Substituents:
VSEPR Angle? Polar?	VSEPR Angle? Polar?

Sketch Observations for Three Students (Three Substituents):

Three Identical Substituents:	Three Different Substituents:
VSEPR Angle? Polar?	VSEPR Angle? Polar?

Sketch Observations for Four Students (Four Substituents):

Four Identical Substituents:	Four Different Substituents:
VSEPR Angle? Polar?	VSEPR Angle? Polar?

Continued on Back →

Questions:

1) Using the crossed-arrow notation, show how the polar bonds in CO_2 cancel out to produce a nonpolar molecule and the polar bonds in H_2O do not cancel out. Be sure to use the correct VSEPR bond angles.

2) What simple, general rule can you come up with to help you determine when polar bonds will cancel each other out, resulting in a nonpolar molecule?

3) Based on what we have seen in this activity, what are the two general rules for predicting trends in boiling point?

a) In general, when given two molecules with similar numbers of electrons, if one is polar and the other is nonpolar...

b) In general, when given two molecules with different numbers of electrons, if both are polar or both are nonpolar...

4) Complete the following table. For each pair of molecules circle the one you predict will have the higher boiling point. Your answer should be based on the number of electrons in each molecule, the polarity of each molecule, and the application of the rules you came up with in Question #3.

	Molecules	# of Electrons	Polar?		Type of IMF?	
Pair #1	CH_4		Yes	No	Induced	Permanent
	SiH_4		Yes	No	Induced	Permanent
Pair #2	CH_2Cl_2		Yes	No	Induced	Permanent
	CF_4		Yes	No	Induced	Permanent
Pair #3	AsH_3		Yes	No	Induced	Permanent
	PH_3		Yes	No	Induced	Permanent
Pair #4	CH_4		Yes	No	Induced	Permanent
	CH_3CH_3		Yes	No	Induced	Permanent

Part D: An Application of Molecular Polarity

Instructions: Add 20 mL of <u>water</u> and a few drops of <u>food coloring</u> to a <u>resealable plastic baggie</u>. After mixing the water and food coloring, add 20 mL of <u>vegetable oil</u>. Remove as much air from the baggie as possible and seal it. Shake the baggie for 10 seconds and record your observations. Let the baggie rest for 2 minutes and again record your observations. Open the bag and add a few squirts of <u>liquid soap</u>. Remove as much air as possible and reseal the bag. Shake the baggie for 10 seconds and record your observations. Again, let the baggie rest for 2 minutes and record your observations

Observations:
Immediately After I Stopped Shaking the Bag (No Soap):

About 2 Minutes After I Stopped Shaking the Bag (No Soap):

Immediately After I Stopped Shaking the Bag (with Soap):

About 2 Minutes After I Stopped Shaking the Bag (with Soap):

Questions:
1) Which liquid is more dense—oil or water? Explain.

2) In general, only polar substances mix with other polar substances, and only nonpolar substances mix with other nonpolar substances. Based on this, is the food coloring polar or nonpolar? Explain.

3) Is the vegetable oil polar or nonpolar? Explain.

4) In this experiment, the soap is acting as a bridge between the water and the oil. What does that tell you about the nature of the soap molecule?

Name: _____   Score: _____   points

1) The high price of a barrel of crude oil is constantly in the news. Crude oil is a mixture containing many different compounds, for example, propane (C_3H_8, used for stoves) and octane (C_8H_{18}, used for automobile gasoline) and dodecane ($C_{12}H_{26}$, used as fuel for jet engines and tractors). To be useful, the various chemicals in crude oil must be separated from each other using their boiling points.

a) Draw the Lewis dot structure for propane.

b) What type of IMF is present in liquid propane? Explain.

c) Imagine you have a sample of oil with only C_3H_8, C_8H_{18}, and $C_{12}H_{26}$. As you start heating the liquid, which compound would you expect to boil first? Last? Explain your answers.

d) Microwave ovens work by causing polar molecules (for example, the water in food) to rotate. The food is heated as the water molecules are caused to spin around and bump into each other. Could a microwave oven be used to heat a sample of crude oil? Explain.

Continued on Back →

2) Using the crossed-arrow notation, show how the polar bonds in CH_4 cancel out to produce a nonpolar molecule. Then do the same showing that the polar bonds in NH_3 do not cancel out to produce a polar molecule. Be sure to use the correct VSEPR bond angles.

3) Which of the two compounds in Question #2 is expected to have a higher boiling point? Explain.

4) Vitamin A is fat-soluble and can be stored in the body, whereas vitamin C is water-soluble and is readily excreted from the body with urine. For this reason, we need frequent intake of vitamin C. Knowing that body fat is an oily (and therefore, nonpolar) material, which of the two vitamins is polar? Which is nonpolar? Briefly explain your answer.

5) Explain what is meant by the following statement: "When considering what happens to the shared bonding electrons when a compound is formed, we should realize that there is a continuum with nonpolar covalent bonds at one extreme, ionic bonds at the other extreme, and polar covalent bonds somewhere in the middle." Use NaCl, ICl, and Cl_2 as part of your explanation.

Activity 6d: The Amazing Properties of Water
COVER SHEET

Before Class

- Read the pages in your textbook dealing with "hydrogen bonding" and "properties of water." Your instructor may assign specific pages for you to read.
- Complete the Pre-Activity Problem Set on the next page.

Key Concepts

- Life as we know it cannot exist without water.
- Hydrogen bonding is an unusually strong type of intermolecular force between molecules with permanent dipoles.
- Hydrogen bonding is responsible for a wide variety of water's unusual properties.

Learning Objectives

- You will be able to define hydrogen bonding and identify when it can occur.
- You will be able to explain some everyday occurrences in terms of the properties of water.
- You will be able to explain the important properties of water in terms of the behavior of the water molecules.

Pre-Activity Problem Set: Activity 6d

Name: **Score:** **points**

1) Use your textbook or some other resource to define the terms (a – d) below:
   a) *Capillary action*

   b) *Viscosity*

   c) *Adhesion/Cohesion*

   d) *Surface tension*

2) Draw the three-dimensional VSEPR structure (with labeled bond angles) for water. Clearly indicate how the VSEPR structure supports the idea that water is a polar molecule.

3) What is unusual about the density of ice versus that of water? How does the density of ice compared to water allow life to exist in lakes during the winter freeze?

Continued on Back →

4) What is meant by the term *hydrogen bonding*? What are the specific requirements for hydrogen bonding? Hydrogen bonding is considered to be a special case of what type of intermolecular force?

5) Using small filled circles to represent hydrogen atoms (●) and slightly larger empty circles (○) to represent oxygen atoms, draw a submicroscopic representation of water at 110°C (and normal atmospheric pressure) in the box to the right.

6) Are the water molecules in Question #5 moving fast enough to overcome their *intermolecular forces*? Explain.

7) Are the water molecules in Question #5 moving fast enough to overcome their *intramolecular bonds*? Explain.

8) In which of the following cases does the dashed line represent hydrogen bonding between two neighboring HF molecules? Briefly explain your choice.

Explorations in Conceptual Chemistry: Activity 6d

Instructions: Complete the following table, indicating whether each compound is polar or nonpolar, the type of intermolecular forces that would be present (induced or permanent dipole), and the total number of electrons in the compound. When you are done, plot the boiling points (measured at normal atmospheric pressure) for CH_4, SiH_4, and GeH_4 on the blank graph below. Label each data point with the compound associated with it and connect the data points (you are NOT drawing a best-fit line in this case). When you are done, answer the questions that follow.

Compound	Boiling Point	Polar?		Type of Dipole		# of Electrons
GeH_4	-89 °C	Yes	No	Permanent	Induced	
SiH_4	-107 °C	Yes	No	Permanent	Induced	
CH_4	-161 °C	Yes	No	Permanent	Induced	
H_2Se	-41 °C	Yes	No	Permanent	Induced	
H_2S	-60 °C	Yes	No	Permanent	Induced	
H_2O		Yes	No			

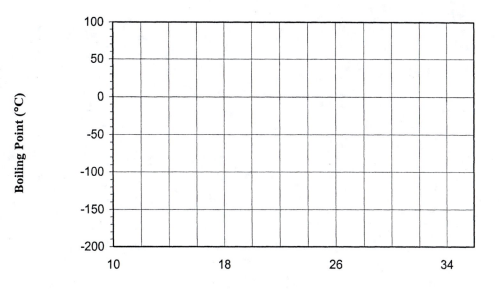

Total # of Electrons

Questions:
1) Using what you have learned about intermolecular forces, explain the trend in boiling points for CH_4, SiH_4, and GeH_4.

Continued on Page 213 →

Add the boiling point data for H_2Se and H_2S to your graph (maybe using a different color pen so it stands out from the first set of data you graphed). Again, label the compounds corresponding to each data point and connect the two points.

2) Compare the boiling point of GeH_4 with that of H_2Se and the boiling point of SiH_4 with that of H_2S. Explain the results of each comparison.

Based on the trend for CH_4, SiH_4, and GeH_4, place a small "x" on your graph where you would *expect* the boiling point of H_2O to be. Now draw and label the data point for what you know to be the actual boiling point of water. Connect this actual data point to the line for H_2Se and H_2S.

3) What is the specific name given to the type of intermolecular force between two H_2O molecules?

4) Which intermolecular force is stronger, the one between two H_2S molecules or the one between two H_2O molecules? Fully explain your answer using the boiling points for the two compounds.

5) How do you explain the extreme difference between the expected and actual boiling points for H_2O?

6) Draw and clearly label a picture showing the intermolecular force between two water molecules. Clearly label the covalent bonds and the intermolecular force.

Instructions #1: Insert the open end of a <u>capillary tube (a small, hollow glass tube that is closed at one end</u> into a <u>small dish or beaker</u> of <u>water</u>. Draw a side view of your observations below. Repeat the test with a small dish of <u>vegetable oil</u> in place of the water. Dispose of the vegetable oil according to your instructor's directions. Clean the glassware with soap and water.

Observations (Side-View Drawings):

Water: Oil:

Questions:

1) Which property/properties from your Pre-Activity Problem Set (Question #1) is being observed in the glass tube test? Explain.

2) Do you think that the surface of glass is polar or nonpolar? Explain.

Instructions #2: Drop a <u>small ball bearing</u> into a <u>large test tube or graduated cylinder</u> three-fourths filled with <u>water</u>. Use a <u>stopwatch</u> to record the time it takes the ball bearing to reach the bottom. Repeat the measurement two more times and find the average of the three trials. If you lose a lot of water removing the ball bearing, be sure to bring the level back up to three-fourths. Next, perform three trials to see how long the same ball bearing takes to reach the bottom of a test tube filled with <u>honey (or another thick liquid)</u>. Dispose of the honey according to your instructor's directions. Clean the glassware with soap and water.

Observations:

	Trial #1	Trial #2	Trial #3	Average
Water				
Honey				

Questions:

1) Which property/properties from your Pre-Activity Problem Set (Question #1) is being measured in the ball bearing drop test? Explain.

2) Why do you think it is typical for scientists to perform multiple trials for important measurements or results?

Continued on Next Page →

Instructions #3: Thoroughly wet about <u>two feet of yarn (not string)</u> with <u>water</u>. Set <u>two small beakers</u> on your desk approximately 1 foot apart. Half-fill one beaker with water and add a few drops of <u>food coloring</u>. Run the wet yarn between the rims of the two beakers, with the excess yarn hanging into the beakers. Lift the beaker with the water about 1 foot above the desk. Keeping the yarn taut and holding it in place, slowly tip the beaker and allow the water to run along the yarn to the empty beaker sitting on your desk.

Observations:

Questions:
1) Which property/properties from your Pre-Activity Problem Set (Question #1) is being observed in the yarn test? Explain.

2) Fully explain in terms of the behavior of the water molecules, why the water flows down the string into the other beaker rather than spilling to the desk.

Instructions #4: Get a <u>small beaker</u> of <u>water</u> and a <u>penny</u>. Record your prediction for the number of drops of water you think you can place on the penny. Use an <u>eyedropper or disposable pipette</u> to place as many drops of water as you can on the penny without spilling over. When you think you are getting close to spilling, draw a side view of the penny and the water. Record the final number of drops you were able to add before spilling.

Observations:
Predicted Number of Drops: Actual Number of Drops:

Side-View Drawing:

Questions:
1) Which property/properties from your Pre-Activity Problem Set (Question #1) is being measured in the penny test? Explain.

2) Why does the water drop on the penny take on the shape that it does?

Name: **Score:**   **points**

1) Match each of the following observations (below left), with one of the corresponding properties of water (below right):

    a) A leaf floats on the water.   • Low viscosity

    b) A towel soaks up water.   • Low density of ice

    c) Water flows readily in rivers and streams.   • Capillary action

    d) A bottle of water breaks open in the freezer.   • Surface tension

2) Referring to Question #1d above, use your textbook to explain what is happening to the water molecules that causes a macroscopic sample of water to expand when it freezes.

3) You may know that if you heat a pot of water on the stove, the temperature of the water will continue to increase until the water begins to boil. At that point, the temperature of the water stops increasing until all the water has turned to gas. Why doesn't the temperature of the water increase while it is boiling? In other words, what does all the heat that is being added to the water do if it isn't speeding up the water molecules? Briefly explain.

4) Try the following experiment at home and then answer the two questions below.
    Instructions: Fill a <u>shallow dish or pan</u> with <u>water</u>. Allow the dish to sit for a minute so that the water is still. Sprinkle the entire surface of the water with <u>pepper</u>. Carefully add a single drop of <u>liquid soap</u> into the center of the pan.
    a) What did the pepper do when you sprinkled it on the water? What property of water does the pepper allow you to observe? Briefly explain.

    b) What did the pepper do when you added the drop of soap? Briefly explain your observation in terms of the properties of soap molecules.

Activity 6e: Symbolic Representations of Molecules
COVER SHEET

Before Class
- Read the pages in your textbook dealing with "organic chemistry." Your instructor may assign specific pages for you to read.
- Complete the Pre-Activity Problem Set on the next page. You will need to use the Internet for some parts of the Pre-Activity Problem Set.

Key Concepts
- Depending on the type of information being conveyed, scientists use various forms of symbolic representations when depicting molecules.
- Functional groups are combinations of atoms that behave as a unit and that impart specific properties to a molecule.
- Ignorance is dangerous!

Learning Objectives
- You will be able to name binary diatomic molecules.
- You will be able to identify basic functional groups.
- You will be able to convert between condensed, structural, skeletal, and perspective formulas.

Name: _____   Score: _____ points

1) What is an isomer? Draw the two isomers for butane, C_4H_{10}.

2) The following questions refer to organic chemistry functional groups.

a) Draw and name the functional group that is required for a compound to be called aromatic.	b) Draw and name the functional group that is required for a compound to be an alcohol.

3) Find out about the highly controversial compound dihydrogen monoxide at the official website: http://www.dhmo.org/. Feel free to look around the website, but be sure to check out the "FAQs" to help you answer the following questions:

    a) List three health-environmental-related problems associated with dihydrogen monoxide.

        i)

        ii)

        iii)

    b) List three uses of dihydrogen monoxide.

        i)

        ii)

        iii)

    c) Based on an analysis of benefits versus risks, would you be in support of a ban of dihydrogen monoxide? Explain.

Instructions: Use the following rules to name the molecules at the bottom of the page. **NOTE:** These rules do NOT apply to ionic compounds (i.e., salts that contain a metal and a nonmetal).

1) The elements are named in the order they appear in the formula (usually this means the element farthest to the left on the periodic table is named first).

2) The first element in the compound is given its normal name.

3) The ending of the second element in the name is changed to *-ide*.

4) Greek prefixes are used to indicate the number of atoms of each element present (see the table below).

5) With regard to the prefixes:
   a) *Mono* is only used for the second element.
   b) The "o" is left off the prefix when the name of the element begins with a vowel.

Prefix	Meaning
mono-	1
di-	2
tri-	3
tetra-	4
penta-	5
hexa-	6

Apply the above rules to name the following covalent compounds:

Formula	Name
CO	
CO_2	
N_2O_4	
Cl_2O	
CCl_4	

Instructions: Your instructor will briefly discuss the following symbolic representations of butane.

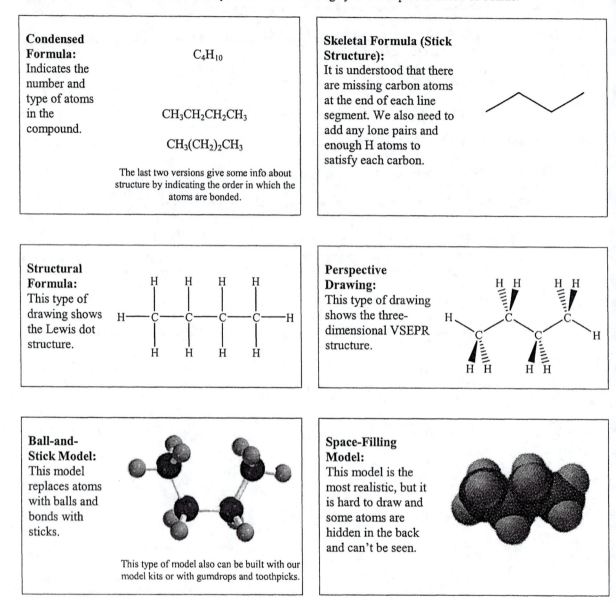

Condensed Formula: Indicates the number and type of atoms in the compound.

$$C_4H_{10}$$

$$CH_3CH_2CH_2CH_3$$

$$CH_3(CH_2)_2CH_3$$

The last two versions give some info about structure by indicating the order in which the atoms are bonded.

Skeletal Formula (Stick Structure): It is understood that there are missing carbon atoms at the end of each line segment. We also need to add any lone pairs and enough H atoms to satisfy each carbon.

Structural Formula: This type of drawing shows the Lewis dot structure.

Perspective Drawing: This type of drawing shows the three-dimensional VSEPR structure.

Ball-and-Stick Model: This model replaces atoms with balls and bonds with sticks.

This type of model also can be built with our model kits or with gumdrops and toothpicks.

Space-Filling Model: This model is the most realistic, but it is hard to draw and some atoms are hidden in the back and can't be seen.

This is the skeletal formula for the aspirin molecule. Follow along in the space below as your instructor converts the skeletal formula to the structural and condensed formulas. Identify the bond angles #1 and #2 in the skeletal formula.

Skeletal Formula:	Structural Formula:
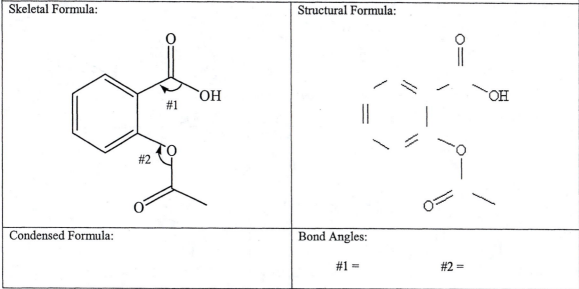	
Condensed Formula:	**Bond Angles:**
	#1 = #2 =

This is the skeletal formula for capsaicin. Cayenne peppers get their heat from this "spicy" molecule. Draw the structural and condensed formulas for capsaicin in the space below. Identify the bond angles #1 and #2 in the skeletal formula.

Skeletal Formula:

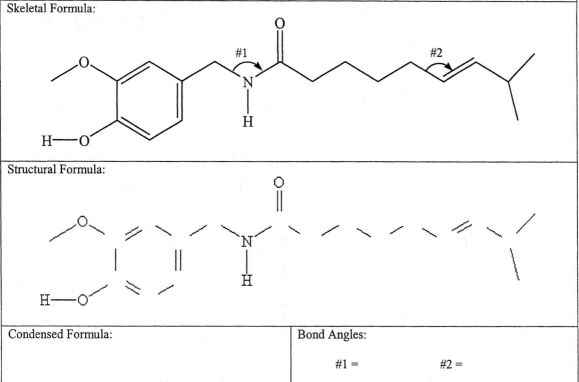

Structural Formula:

Condensed Formula:	Bond Angles:
	#1 = #2 =

Questions:

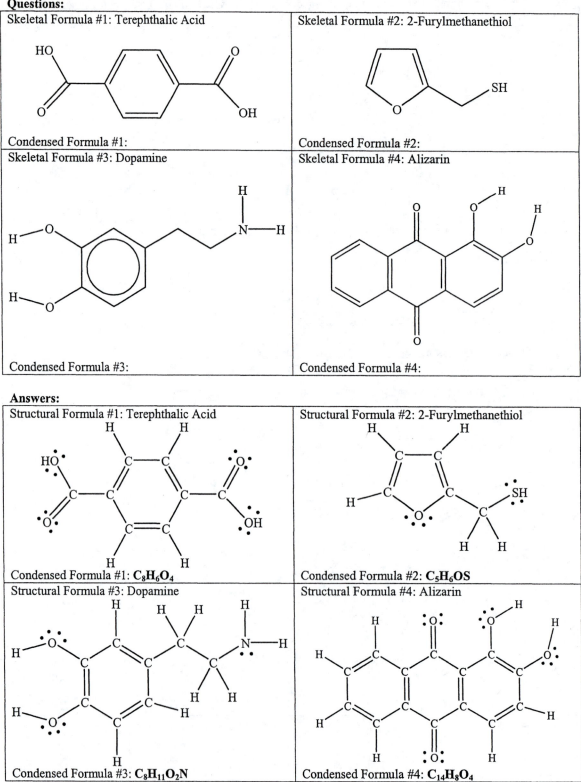

Skeletal Formula #1: Terephthalic Acid

Condensed Formula #1:

Skeletal Formula #2: 2-Furylmethanethiol

Condensed Formula #2:

Skeletal Formula #3: Dopamine

Condensed Formula #3:

Skeletal Formula #4: Alizarin

Condensed Formula #4:

Answers:

Structural Formula #1: Terephthalic Acid

Condensed Formula #1: $C_8H_6O_4$

Structural Formula #2: 2-Furylmethanethiol

Condensed Formula #2: C_5H_6OS

Structural Formula #3: Dopamine

Condensed Formula #3: $C_8H_{11}O_2N$

Structural Formula #4: Alizarin

Condensed Formula #4: $C_{14}H_8O_4$

Name: _____ **Score:** _____ **points**

1) Skimming your textbook, choose any molecule with at least six carbon atoms and answer the questions below. When you are done, circle and name at least one of the functional groups found in your molecule.

   a) Compound Name:

   b) Condensed Formula:

   c) Use/Importance of This Compound:

   d) Skeletal Formula:

   e) Structural Formula:

Continued on Back →

2) Check out the "Molecule of the Week" at the American Chemical Society. Go to: http://portal.acs.org/portal/acs/corg/content. If you can't find the "Molecule of the Week" on the main page, use the search box to find it. Once you have found the "Molecule of the Week" page, click on "Archive." After selecting a molecule of your choice, answer the questions below and circle/name at least one of the functional groups found in your molecule.

a) Compound Name:

b) Condensed Formula:

c) Use/Importance of This Compound:

d) Skeletal Formula:

e) Structural Formula:

Activity 6f: An Introduction to Polymers
COVER SHEET

Before Class
- Read the pages in your textbook dealing with "polymers." Your instructor may assign specific pages for you to read.
- Complete the Pre-Activity Problem Set on the next page. You will need to use the Internet for some parts of the Pre-Activity Problem Set.

Key Concepts
- Polymers are long molecules made up of repeating small units called monomers.
- The desired properties of an object dictate what type of polymer it should be made from.
- Polymers are everywhere.

Learning Objectives
- You will be able to discuss the relationship between the uses of an object and the properties of the polymer that it is made from.
- You will be able to determine the monomer used to make a given polymer and vice versa.
- You will be able to perform tests to distinguish between common polymers.

Name: _____   Score: _____   points

1) How is a polymer like a long string of paper clips linked together?

2) The following questions refer to the polymer polyethylene and its monomer ethylene.

   a) Draw the VSEPR structure (with labeled bond angles) for ethylene.

   b) Draw the shorthand representation (showing a single repeat unit) of polyethylene. Refer to Question #3 on the back of this page for an example.

   c) Where does ethylene come from?

   d) What happens to the electrons in the double bond when ethylene monomers are made into polyethylene?

   e) Are the attractive forces *between* polyethylene chains due to permanent dipoles or induced dipoles? Explain.

   f) Explain the difference in structure between high-density polyethylene (HDPE) and low-density polyethylene (LDPE). As part of your answer, draw a representation showing the difference in structure between HDPE and LDPE.

   g) What are some of the characteristic properties of HDPE? What are some of its uses?

   h) What are some of the characteristic properties of LDPE? What are some of its uses?

Continued on Back →

3) Draw the complete VSEPR structure (including all bonds, bond angles, and lone pairs) for the monomer used to make polyacrylonitrile (shown below in shorthand notation).

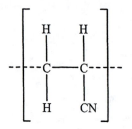

4) Draw the shorthand notation for the polymer formed from the propylene monomer (shown below).

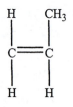

5) Find a Website that identifies the recycling codes for the various polymers. Attach a copy of the codes to your Pre-Activity Problem Set. Find two objects in your house with recycling codes. What are the objects? What are their recycling codes? From what polymers are they made?

Object #1:

Recycling Code: Polymer Name:

Object #2:

Recycling Code: Polymer Name:

Instructions: Your instructor has collected a number of objects made from various polymers. Select three objects (at least one of which has a recycling code on it) and answer the questions below.

Questions:
1) Which three objects did you pick? For each object, discuss how the properties of the polymer are related to the function/use of the object.

Object #1:

Object #2:

Object #3:

2) Which of the three objects has a recycling code? Use the table of recycling codes you found as part of your Pre-Activity Problem Set to identify the polymer used to make the object.

Object:

Recycling Code: Polymer Name:

CAUTION: Avoid inhaling or ingesting dry Borax. There is typically no danger in handling this polymer, but you should wash you hands after contact. Gluep does not typically stick to clothes or walls, but it can stain wooden furniture.

Instructions: In a <u>small beaker</u>, mix 2 <u>teaspoons</u> of <u>Borax solution</u> (made by completely dissolving 1/4 cup of borax in 1 L of water), 1 <u>tablespoon</u> of <u>water</u>, a few drops of <u>food coloring</u> (optional), and 1 tablespoon of <u>Elmer's glue</u>. Stir until the mass begins to stick to the <u>spoon</u>. Remove the gluep from the beaker and work it in your palm. Explore the properties of the polymer you made. Some suggested tests are listed below. Record your tests and your observations. When you are done, either discard the gluep in a garbage can or store it in a <u>plastic baggie</u> for later (a few drops of <u>Lysol</u> can be added to the Borax solution to retard mold growth).

Some Suggested Tests:
- Try stretching the gluep first quickly and then slowly.
- Try squeezing the gluep or pushing your finger into it.
- Write your name backwards with a water-soluble marker. Press the gluep onto the writing to transfer the words.

Observations:

Questions:
1) How were the properties of the Borax solution, the glue, and the water different from the properties of the polymer?

2) Based on its properties, think of a few possible uses for the polymer that you made.

CAUTION: Wear <u>goggles</u> during this part. Calcium chloride can irritate your eyes.

Instructions: Your instructor will supply you with <u>samples of each of the six types of polymers that have</u> <u>recycling codes</u>. Take a sample of each polymer. If they are not color coded, carefully keep track of their unknown identification (A, B, C, D, E, or F). Perform the six tests in the order described below. Record the result of each test in the observation table on page 237. When you are done, return the polymer samples to your instructor to be cleaned and reused. Dispose of the blue and yellow test solutions according to your instructor's directions.

Tests:

#1	Place your six polymer samples in a <u>small beaker</u> with about 30 mL of <u>distilled water</u>. Stir the polymers with a <u>glass rod or wooden craft stick</u> to make sure no bubbles are adhering to the polymers. Use the observation table on the next page to record which samples sink and which float.
#2	Use a <u>spoon</u> to scoop out the samples that are floating and place them in a <u>second small beaker</u> with 30 mL of <u>70% isopropyl alcohol</u> (with blue food coloring for easy identification). Stir with a <u>clean glass rod or a new wooden craft stick</u>. Again, record which samples sink and which float.
#3	Scoop out the samples that are floating from the blue alcohol solution. Using a <u>plastic pipette or a teaspoon,</u> add a small amount of water to the blue alcohol solution containing the sinkers and stir. Continue adding small amounts of water and stirring until you see one of the samples **float**. Record your observations.
#4	Go back to the original beaker of water and take out the sinkers. Place the sinkers in a <u>third small beaker</u> with about 30 mL of the <u>concentrated calcium chloride solution</u> (made by mixing 450 mL of distilled water with 2 cups of calcium chloride and dyed yellow). Stir with a <u>clean glass rod or a new wooden craft stick</u>. Record your observations.
#5	Using a plastic pipette or a teaspoon, add a small amount of water to the yellow calcium chloride solution and stir. Continue adding small amounts of water and stirring until you see one of the samples **sink**. Record your observations.
#6	Continue to add small amounts of water until another one of the samples **sinks**. Record your observations.

Continued on Page 237 →

Observations:

After **Test #1**	⇒	Float:	Sink:
After **Test #2**	⇒	Float:	Sink:
After **Test #3**	⇒	Float:	Sink:
After **Test #4**	⇒	Float:	Sink:
After **Test #5**	⇒	Float:	Sink:
After **Test #6**	⇒	Float:	Sink:

Questions:

1) Imagine you work in a plastic recycling factory where used plastic containers are chopped up into small pieces that are then melted and made into pellets. These pellets can then be used to make things like garbage cans, carpeting, and plastic lumber. Unfortunately, someone forgot to record the identity of the polymers after they were chopped up. To continue with the recycling process, you need to know the identity of each sample.

   a) Using your test results, begin by ranking your polymer samples, identified by color or unknown identification (A, B, C, D, E, or F) from least to most dense. Briefly explain your ranking. For example, "Sample X was the least dense because it always floated."

   Least Dense   Most Dense

   b) Use the density table below to identify each of your polymer samples.

Color (or Unknown Identification Code)	Polymer	Density (g/cm^3)
	Polypropylene (PP)	0.90 – 0.91
	Low-density polyethylene (LDPE)	0.92 – 0.94
	High-density polyethylene (HDPE)	0.95 – 0.97
	Polystyrene (PS)	1.05 – 1.07
	Polyvinyl chloride (PVC or V)	1.16 – 1.35
	Polyethylene terephthalate (PETE or PET)	1.38 – 1.39

Name: _____   Score: _____   points

1) Following the example in the first row, complete the other three rows by listing four properties required of the polymer used to make the object.

Object	Desired properties
Contact lens	Examples: • Transparent • Nontoxic • Semipermeable (to oxygen) • Flexible
Car tire	• • • •
Cling wrap	• • • •
Clothing fiber	• • • •

2) A typical clear plastic water or soda bottle may contain as many as four to five different polymers. How many you can find? In the space below, list each different part of the bottle that you think is made of a different polymer.

Continued on Back →

3) Research the role of plasticizers in polymer manufacturing and answer the questions below. Indicate all the sources of your information.

- What are plasticizers?

- What function do they serve in polymer manufacturing?

- What is the most common class of plasticizers?

- What is the controversy surrounding the use of this class of plasticizers? List the arguments made both for and against their use.

Unit 7: Naming and Exploring Ionic Compounds (Salts)

Ionic compounds (also called "salts" by chemists) are all around us: they allow our muscles and nerves to work, they enhance the flavor of our foods, and they can even help melt ice on snowy roads! In Unit 7, you will learn how to name ionic compounds and will explore some of their properties.

Activity 7: Naming and Exploring Ionic Compounds
COVER SHEET

Before Class

- Read the pages in your textbook dealing with "ionic compounds." Your instructor may assign specific pages for you to read.
- Complete the Pre-Activity Problem Set on the next page.
- Cut out the three pages of puzzle pieces found on pages 249 to 253. Put them in an envelope for safekeeping and bring them to the lab.

- Prepare the Bingo game board according to directions in *Part A* of this Activity.

Key Concepts

- The names used for ionic compounds follow basic rules.
- Some ionic compounds are soluble in water. When added to water, these compounds break up to give ions.
- Some ionic compounds are insoluble in water. When added to water, these compounds do not generate ions.

Learning Objectives

- You will determine the rules used for naming basic ionic compounds.
- You will be able to apply the rules to name ionic compounds given an ionic formula.
- You will be able to write the formula given the name of an ionic compound.
- You will be able to draw submicroscopic representations of various salts added to water.

Explorations in Conceptual Chemistry: Activity 7   243

Name: **Score:**   **points**

Instructions: Use the following examples to figure out the rules for naming ionic compounds. Feel free to refer to a periodic table and the handout "Names of Some Common Positive and Negative Ions" found on page 247.

Sodium chloride is the most common example of an ionic compound and is typically referred to as "table salt" or just "salt." You should know, however, that to a chemist a "salt" is *any* ionic compound, not just sodium chloride. Starting with sodium chloride, let's see what we can figure out about the rules that are used to name ionic compounds.

Example #1: Basic Rules

NaCl is called sodium chloride.

1a) Since we are learning to name ionic compounds, there must be a metal and a nonmetal in the formula. What is the metal?   What is the nonmetal?

1b) Where is the metal written in the formula?

1c) Where is the nonmetal written in the formula?

1d) What happened to the name of the metal when it went from being a pure element to being a part of an ionic compound?

1e) What happened to the name of the nonmetal when it went from being a pure element to being a part of an ionic compound?

Example #2: Balancing Charges

MgI₂ is called magnesium iodide.

2a) Does the name of this compound follow the rules determined in Example #1? Explain.

2b) What is the charge on the magnesium ion?

2c) What does the charge on the magnesium ion have to do with the number of valence electrons that the neutral element magnesium has?

2d) What is the charge on the iodide ion?

2e) What does the charge on the iodide ion have to do with the number of valence electrons that the element iodine has?

2f) How many iodide ions are needed to cancel out the charge on a magnesium ion?

2g) How is the number of iodide ions needed in Question #2f indicated in the formula?

Continued on Back →

Example #3: Metals with Variable Charges
These rules are used with transition metals and elements in the lower part of groups 13 and 14 (for example, Sn and Pb).

CuCl is called copper(I) chloride.
CuCl₂ is called copper(II) chloride.

3a) What is the charge on the chloride ion?

3b) Knowing the charge on the chloride ion, what is the charge on the copper ion in the first compound?

3c) How is the charge on the copper ion indicated in the name of the first compound?

3d) Knowing the charge on the chloride ion, what is the charge on the copper ion in the second compound?

3e) How is the charge on the copper ion indicated in the name of the second compound?

3f) Why don't we need to do this with group 1 and group 2 elements (and aluminum, Al)?

Example #4: Polyatomic Ions

Na₂CO₃ is called sodium carbonate
Mg(NO₂)₂ is called magnesium nitrite
PbCO₃ is called lead(II) carbonate

4a) What do we do to the name of the polyatomic ion when we include it in an ionic compound?

4b) When do we use parentheses in writing the formula for an ionic compound? Your answer should explain why $Mg(NO_2)_2$ has parentheses, while Na_2CO_3 doesn't.

Names of Some Common Positive and Negative Ions

The Names of Some Common Positive Ions (Cations)

Formula	Name	Formula	Name
Li^+	lithium	Fe^{2+}	iron (II)
Na^+	sodium	Fe^{3+}	iron (III)
K^+	potassium	Cu^+	copper (I)
Rb^+	rubidium	Cu^{2+}	copper (II)
		Sn^{2+}	tin (II)
Mg^{2+}	magnesium	Sn^{4+}	tin (IV)
Ca^{2+}	calcium	Pb^{2+}	lead (II)
Sr^{2+}	strontium	Pb^{4+}	lead (IV)
Al^{3+}	aluminum	NH_4^+	ammonium

The Names of Some Common Monatomic Negative Ions (Anions)

Formula	Name	Formula	Name
Cl^-	chloride	O^{2-}	oxide
F^-	fluoride	S^{2-}	sulfide
Br^-	bromide		
I^-	iodide	N^{3-}	nitride

The Names of Some Common Polyatomic Negative Ions (Anions)

Formula	Name	Formula	Name
$C_2H_3O_2^-$	acetate	CO_3^{2-}	carbonate
CN^-	cyanide	SO_4^{2-}	sulfate
OH^-	hydroxide	SO_3^{2-}	sulfite
NO_3^-	nitrate	CrO_4^{2-}	chromate
NO_2^-	nitrite	$Cr_2O_7^{2-}$	dichromate
ClO_4^-	perchlorate	HPO_4^{2-}	hydrogen phosphate
ClO^-	hypochlorite		
MnO_4^-	permanganate	PO_4^{3-}	phosphate
HCO_3^-	hydrogen carbonate (or bicarbonate)		

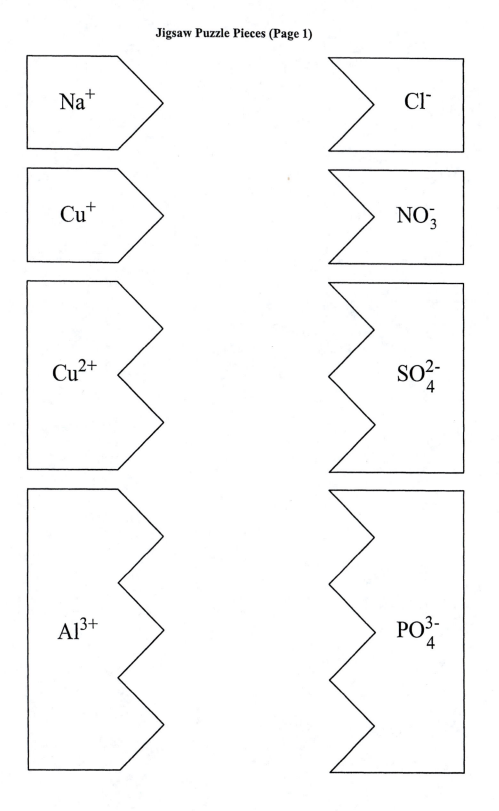

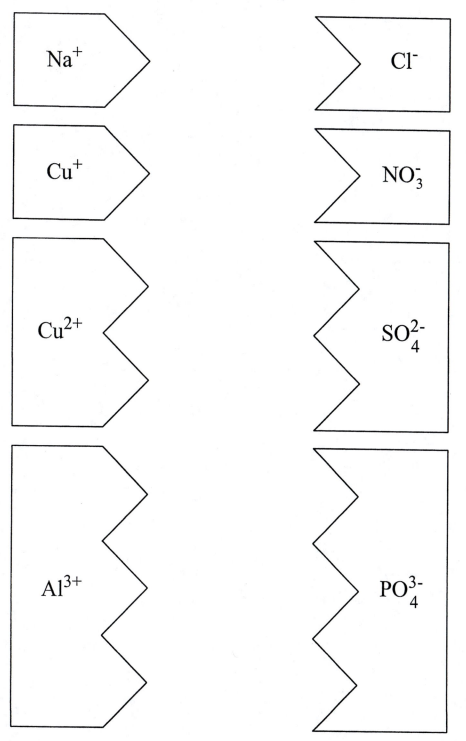

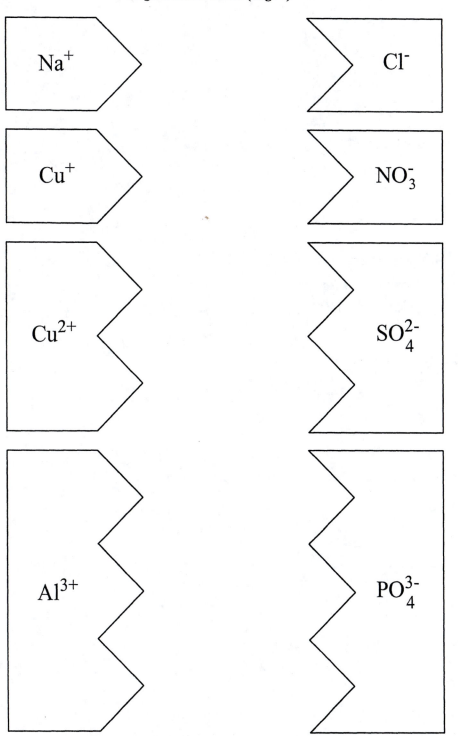

Na^+

Cl^-

Cu^+

NO_3^-

Cu^{2+}

SO_4^{2-}

Al^{3+}

PO_4^{3-}

Before Class: Remove the handout "Names of Some Common Positive and Negative Ions" from your activity manual. Select any combination of 24 names or formulas from the handout and write one in each of the Bingo squares (except the "FREE" square) at the bottom of this page. For example, you might write the formula "K^+" in one square and the name "sodium" in another and so on. You will use this completed Bingo game board to play in class.

Instructions: If your instructor writes the name of an ion on the blackboard, find the corresponding formula on your game board and mark the square. If your instructor writes a formula on the blackboard, find the corresponding name on your game board and mark the square. For example, if given the name "nitrate," you try to find the formula "NO_3^-" on your game board. If given the formula "SO_4^{2-}," you try to find the name "sulfate" on your game board. The first person to fill in five squares vertically, horizontally, or diagonally yells "Bingo."

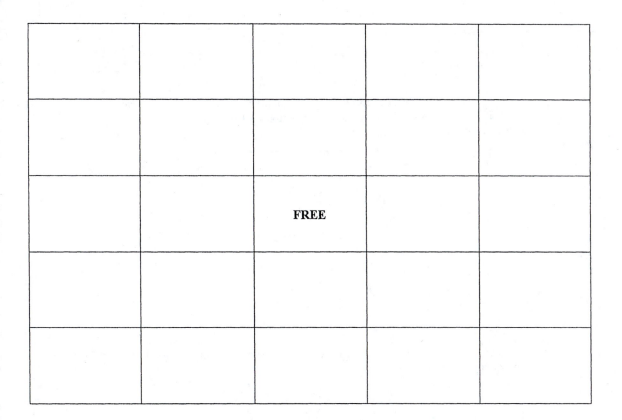

		FREE		

Instructions: Working in pairs, take the jigsaw pieces you cut out for the Pre-Activity Problem Set and form, one at a time, all 16 possible ionic compounds using the puzzle pieces. As you form each possible compound, complete the table on the next page.

Helpful Guidelines to Remember:

*   Note that each puzzle piece has a different number of peaks (cations with positive charges) and valleys (anions with negative charges) depending on the ion charge.

For example:

Each sodium ion has a +1 charge, so the puzzle piece with the sodium ion has one peak:

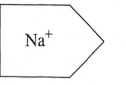

Each sulfate ion has a -2 charge, so the puzzle piece with the sulfate ion has two valleys:

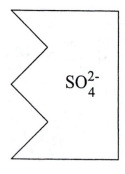

*   Use the correct number of each of the pieces to see in what ratio the ions combine to form the final, neutral compound. A neutral compound has the same total number of positive charges as negative charges and therefore, has no net charge.

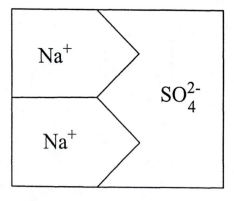

*   Do not form compounds that have more than one different type of cation or more than one different type of anion. An example of what not to do:

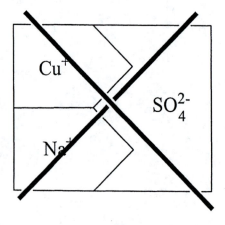

Continued on Next Page →

Question: Following the example below, fill in the table. Indicate the ratio needed to form a neutral ionic compound using each of the possible combinations of anion and cation puzzle pieces. In addition, write the formula and name of each ionic compound.

Cation		Anion		Ratio		Formula		Name
Na^+	combines with	SO_4^{2-}	in a ratio of	2:1	to give	Na_2SO_4	which is called	sodium sulfate

You can check all your answers using the answer key on page 259. Be sure to get help from your instructor or a classmate if you got any wrong!

Explorations in Conceptual Chemistry: Activity 7

Answer Key for Jigsaw Puzzle

Cation		Anion		Ratio		Formula		Name
Na^+		SO_4^{2-}		2:1		Na_2SO_4		sodium sulfate
Na^+		Cl^-		1:1		$NaCl$		sodium chloride
Na^+		NO_3^-		1:1		$NaNO_3$		sodium nitrate
Na^+		PO_4^{3-}		3:1		Na_3PO_4		sodium phosphate
Cu^+	combines with	SO_4^{2-}	in a ratio of	2:1	to give	Cu_2SO_4	which is called	copper(I) sulfate
Cu^+		Cl^-		1:1		$CuCl$		copper(I) chloride
Cu^+		NO_3^-		1:1		$CuNO_3$		copper(I) nitrate
Cu^+		PO_4^{3-}		3:1		Cu_3PO_4		copper(I) phosphate
Cu^{2+}		SO_4^{2-}		1:1		$CuSO_4$		copper(II) sulfate
Cu^{2+}		Cl^-		1:2		$CuCl_2$		copper(II) chloride
Cu^{2+}		NO_3^-		1:2		$Cu(NO_3)_2$		copper(II) nitrate
Cu^{2+}		PO_4^{3-}		3:2		$Cu_3(PO_4)_2$		copper(II) phosphate
Al^{3+}		SO_4^{2-}		2:3		$Al_2(SO_4)_3$		aluminum sulfate
Al^{3+}		Cl^-		1:3		$AlCl_3$		aluminum chloride
Al^{3+}		NO_3^-		1:3		$Al(NO_3)_3$		aluminum nitrate
Al^{3+}		PO_4^{3-}		1:1		$AlPO_4$		aluminum phosphate

Part C: The Ionic Compound Naming Game

Instructions: Playing in groups of three to four students, your group will need <u>one set of "The Ultimate Naming Game" game boards</u> (found on pages 263 and 265). Take turns rolling the <u>die</u>. Each student rolls the die three times. The first roll indicates which set of game boards that student will play with for that turn. The first page of game boards should be used when an EVEN number is rolled and the second page of game boards is used when an ODD number is rolled. The second roll of the die indicates how many spaces to move the <u>place marker</u> on the first game board (CATIONS). The die is then rolled a third time indicating how many spaces to move on the second board (ANIONS). After each person's turn, write the name and formula rolled in the table below.

- *When playing with the EVEN-number boards*, **each student** should write down the name and formula for the resulting ionic compound that is formed. Compare answers with the others in your group.
- *When playing with the ODD-number boards*, **each student** should write down the name and formula of the resulting ionic compound. When naming compounds, remember to consider when roman numerals are needed. Compare answers with the others in your group.

Name	Formula		Name	Formula

The Ultimate Naming Game (*Part C*):

Use the boards on this page for an **EVEN** roll of the die.

START → tin (IV)	lithium	iron (II)	lead (II)
calcium	**CATIONS** (positive ions)		aluminum
rubidium			lead (IV)
copper (II)			tin (II)
magnesium			copper (I)
iron (III)	strontium	potassium	sodium

START → fluoride	chloride	sulfite	perchlorate
hydrogen carbonate	**ANIONS** (negative ions)		oxide
acetate			nitride
phosphate			hydroxide
carbonate			dichromate
sulfide	cyanide	nitrate	sulfate

The Ultimate Naming Game (*Part C*).

Use the boards on this page for an **ODD** roll of the die.

START → Ca^{2+}	Li^+	Fe^{2+}	Cu^+
Pb^{2+}			Sr^{2+}
Mg^{2+}	**CATIONS** (positive ions)		Al^{3+}
Ba^{2+}			Pb^{4+}
K^+			Fe^{3+}
Na^+	Rb^+	Cu^{2+}	Sn^{2+}

START → Cl^-	F^-	Br^-	NO_3^-
S^{2-}			I^-
CN^-	**ANIONS** (negative ions)		N^{3-}
OH^-			O^{2-}
$C_2H_3O_2^-$			MnO_4^-
SO_4^{2-}	CO_3^{2-}	PO_4^{3-}	HCO_3^-

Instructions: Half-fill a <u>styrofoam cup or beaker</u> with <u>ice</u>. Use a <u>thermometer</u> to record the initial temperature of the ice. Add a <u>teaspoon</u> of <u>sodium chloride</u> (table salt) to the cup. Stir until the salt is dissolved. Record the final, lowest temperature of the ice and salt.

Observations:

Temperature of Ice: Temperature of Ice and Salt:

Questions:

1) What happened to the freezing point of the ice when the salt was added? How does this observation explain why salt is sometimes added to icy roads during a snow storm?

2) Not all ionic compounds dissolve in water; for example, if you put a scoop of barium sulfate into water, it will just sit on the bottom as a solid. If they do dissolve in water, however, ionic compounds always break up into their individual ions. It is these ions (mobile, charged particles) that allow salts to conduct electricity when dissolved in water. Knowing that sodium acetate and iron(III) nitrate will dissolve in water and that barium sulfate will not, draw submicroscopic representations of these three ionic compounds when added to water. For clarity reasons, don't draw the water molecules.

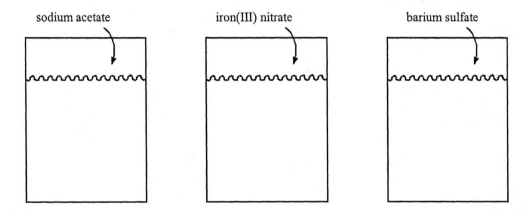

3) Most of us have heard the term *electrolytes* in reference to biological health. An electrolyte is the scientific term for compounds that form ions when added to water. In our bodies, the carefully controlled differences in the amount of these ions inside and outside of the cell membrane are responsible for important cell functions, such as the electrical impulses in nerve cells and the muscle contraction in muscle cells. Replenishing electrolytes lost during heavy sweating is the basis behind sports drinks. The important electrolytes in our bodies include Na^+, K^+, Ca^{2+}, Mg^{2+}, Cl^-, HCO_3^-, SO_4^{2-}, and PO_4^{3-}. Name these biologically important electrolytes.

Name: _____   Score: _____ points

1) A bottle of Soft Scrub® lists among its contents the compound sodium hypochlorite. What is the formula for this compound?

2) TUMS® antacids contain $CaCO_3$. What is the name of this compound?

3) Answer the following questions.
   a) What is the formula and name for the ionic compound formed from magnesium and oxygen?

   b) What is the formula of aluminum hydrogen phosphate?

   c) What is the name of $Fe_2(CO_3)_3$?

   d) What is the formula for iron(II) carbonate?

4) Given the following formulas, what, if anything, is wrong with their names? Make any appropriate corrections **to the name** so that it matches the given formula.
   a) $Al(NO_3)_3$ aluminum nitrite

   b) $SrSO_4$ strontium(II) sulfate

   c) Ca_3N_2 tricalcium dinitride

   d) Cu_3PO_4 copper(II) phosphate

5) The oxalate ion, $C_2O_4^{2-}$, is an ion that we have not seen yet. Ionic compounds that contain the oxalate ion are named in the same way as the other ionic compounds we have discussed in this activity. Knowing this, answer the following questions.
   a) Write the formula for tin(IV) oxalate.

   b) What is the name for $Na_2C_2O_4$?

Unit 8: Understanding Mixtures and Solutions

Now that you have learned about compounds, we turn our attention to understanding mixtures and solutions. As an introduction, you will separate the components of several mixtures, make observations about the relationship between density and concentration of a solution, and explore the concept of saturation.

After the basics, you will apply what you have learned to several interesting topics. Whether you are a doctor checking the acidity of bodily fluid to diagnose a disease, an environmental scientist interested in the effect of acid rain on the environment, or you are just worried about the acidity of your garden soil, your fish tank, or your swimming pool, it is important to understand the behavior of acids and bases (the topic of Activity 8b). In Activity 8c, you will learn how scientists can use a calibration curve to determine the concentration of a specific solute in a solution. Specifically, you will use a calibration curve to determine the amount of sugar in soda. In the final Activity of Unit 8, you will examine how chemistry is used in making water safe to drink.

Activity 8a: An Introduction to Mixtures and Solutions
COVER SHEET

Before Class
- Read the pages in your textbook dealing with "mixtures and solutions," "concentration," and "saturation." Your instructor may assign specific pages for you to read.
- Complete the Pre-Activity Problem Set on the next page.

Key Concepts
- Most of the matter around us exists as mixtures.
- Mixtures are combinations of two or more substances in which each substance retains its own properties.
- Mixtures can be homogeneous (solutions and suspensions) or heterogeneous.
- The components that make up a mixture can be separated based on differences in their physical properties.
- The concentration of a solution describes the amount of solute dissolved in the solvent.
- Saturated solutions have the maximum amount of a specific solute dissolved in the solvent at the given temperature.

Learning Objectives
- You will be able to identify examples of common mixtures (e.g., air, saltwater, steel, blood, a cup of coffee, etc.).
- You will be able to identify the solute and the solvent in a solution.
- You will be able to decide if a mixture is homogeneous or heterogeneous.
- You will be able to separate the various components of a mixture.
- You will be able to decide which of a series of solutions is more concentrated.

Name: _____   Score: _____ points

1) How can you tell, just by looking, if a mixture is most likely homogeneous or heterogeneous?

2) Classify each of the following as (i) an element, (ii) a compound, (iii) a solution/homogeneous mixture, or (iv) a heterogeneous mixture.

air _____   a cup of coffee _____   pure water _____

chlorine gas _____   a bowl of cereal _____   a diamond _____

a person _____   a bronze statue _____   sugar _____

3) Based on your answer to Question #2, which of the samples should you be able to separate into individual substances based only on physical properties?

4) The tap water in your home leaves behind white deposits after evaporating. Is your tap water a heterogeneous mixture, a solution, a compound, or an element? Explain.

5) A student adds some sugar to a glass of water. Most of the sugar dissolves, but a small amount sits on the bottom.
   a) What is the solvent?   What is the solute?

   b) Is the solution saturated? Explain.

   c) Because some sugar did not dissolve, would it be appropriate to say that sugar is insoluble in water? Explain.

Continued on Back →

6) After making a cup of tea with a new tea bag, your friend asks to reuse the bag to make a second cup of tea.
   a) Which cup of tea would probably be more concentrated? Why?

   b) What observations (use all of your senses) could you make to help you decide which is more concentrated?

7) If different elements are represented with different colored circles, which of the following drawings illustrates a gas phase mixture? All three boxes show two different elements present; why is it that these other boxes are not considered mixtures? Explain your answers.

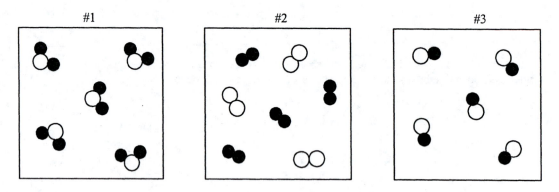

CAUTION: Be careful, hot plates can stay warm for a long time after being turned off. Use heat-protecting gloves when handling hot glassware.

Instructions for Separating Mixture #1: Stir together one heaping <u>teaspoon</u> each of <u>gravel</u>, <u>iron shavings</u>, <u>sand,</u> and <u>salt</u> in a <u>small beaker</u>. Your job is to now develop and carry out your own procedure to separate the four components into separate piles. You will have at your disposal a <u>wire mesh</u>, a <u>coffee filter</u>, a <u>funnel</u>, assorted <u>beakers</u>, a <u>magnet</u>, <u>water</u>, a <u>hot plate</u>, and <u>insulating gloves</u>.

Questions:

1) For each component, what property is being used as the basis for its separation?

   Gravel:

   Salt:

   Iron Shavings:

   Sand:

2) In the space below, write the protocol for the separation of the four-component mixture you performed. Your procedure should be clear enough for another student to follow your steps.

Continued on Back →

Instructions for Separating Mixture #2: You have been hired for your first case as a forensics specialist to help identify the pen used to write a ransom note. Using <u>scissors,</u> cut a piece of <u>filter paper</u> into a 2-cm wide strip. Draw a line 2 cm from the bottom of the strip with a <u>pencil</u>. Using the <u>pen from Suspect #1</u> put a dot (about this big: ●) in the middle of the pencil line. Label this Strip #1. Repeat with the <u>pen from Suspect #2</u>. Label this Strip #2. Take a prepared 2-cm wide <u>strip of the ransom note</u>. Place all three strips (the two test strips and the strip from the ransom note) vertically in a <u>small beaker</u> with a thin layer of <u>water</u> in the bottom. The strips should be touching the water, but the dot should not be in the water. <u>Tape</u> the tops of the strips to the beaker if you are having trouble getting them to stand up. Wait 15 – 20 minutes for the process to be complete. Save your paper strips to attach below.

Observations: Attach your tested ink strips below with the starting end on the left.

Ransom note:

Suspect #1:

Suspect #2:

Questions:
1) What color ink was used to write the ransom note? What colors were you able to separate?

2) Knowing what we do about the nature of water, which of the separated colors from the ransom note is due to molecules that are the most polar (and are attracted to the water as it rises up the paper)? Explain.

3) What color would you expect to get if you remixed the colors that you were able to separate from the ink? Explain.

4) Is the ink used to write the ransom note an element, compound, solution (homogeneous mixture), or heterogeneous mixture? Explain.

5) Which of the two pens was most likely used to write the ransom note? Explain.

Instructions: Begin by making the three solutions described below. Be sure to stir each solution with a <u>glass rod or wooden stirrer</u> until all the salt has dissolved. It may take several minutes of stirring to get all of the salt to dissolve.

- Solution #1: Add 200 mL of <u>water</u>, 5 level <u>teaspoons</u> of <u>salt</u>, and 1 drop of <u>blue food coloring</u> to a <u>400-mL beaker</u>.
- Solution #2: Add 200 mL of <u>water</u>, 8 level <u>teaspoons</u> of <u>salt</u>, and 1 drop of <u>red food coloring</u> to a <u>400-mL beaker</u>.
- Solution #3: Add 500 mL of <u>water</u>, 8 level <u>teaspoons</u> of <u>salt</u>, and 1 drop of <u>yellow food coloring</u> to a <u>600-mL beaker</u>.

Next, assemble your density tester: Attach a <u>wooden stirrer or glass rod</u> to a <u>50-mL beaker</u> using an <u>elastic band</u>. The bottom of the stirrer/rod should be flush with the bottom of the beaker. The long part of the stirrer/rod sticking up will be used as a handle.

Test the relative densities of Solutions #1 and #2 by filling your density tester with Solution #1 and *very slowly* submerging the tester in the beaker containing Solution #2. Record your observations. Clean your density tester and compare Solutions #1 and #3 by filling your density tester with Solution #1 and *very slowly* submerging it in Solution #3. Record your results.

Observations:
Compare Solutions #1 and #2:

Compare Solutions #1 and #3:

Questions:
1) Based on your observations, is Solution #1 or Solution #2 more dense? Why do the relative densities of these two solutions make sense in terms of their composition? Explain your answers.

2) Based on your observations, is Solution #1 or Solution #3 more dense? Why might the relative densities of these two solutions NOT seem to make sense in terms of their composition? Explain your answers.

Continued on Back →

3) The drawings below represent beakers of the three solutions you made in this activity. Use the model drawn for Solution #1 to complete the drawings for Solutions #2 and #3. In the sample drawing, the water level is indicated at 200 mL (each small dash on the beaker indicates 100 mL) and each X represents 1 tsp of solute (NaCl) that was added.

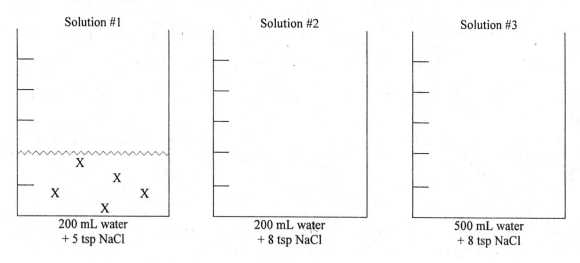

Solution #1	Solution #2	Solution #3
200 mL water	200 mL water	500 mL water
+ 5 tsp NaCl	+ 8 tsp NaCl	+ 8 tsp NaCl

4) How could your three drawings help illustrate why the relative densities of Solutions #1 and #3 turned out the way they did, even though Solution #3 has more salt added to it?

5) What is the important factor when deciding the relative concentrations (and therefore densities) of two solutions made with the same solute and solvent? Explain.

6) How many teaspoons of salt would have to be added to 500 mL of water to give a solution that has the same density/concentration as Solution #1? Show how you determined your answer.

Instructions: For your second case as a forensic scientist you have been asked to help solve a burglary. Three suspects have been identified as the possible thief and, though no fingerprints were left behind, an unknown white powder was found at the crime scene. Searches of the three suspects' homes have each turned up samples of clothing with different white powders that they got on their clothes from work: Suspect #1 works at a plant that processes Epsom salt, Suspect #2 works in a plant that processes calcium carbonate, and Suspect #3 works at a plant that packages sodium chloride. Your job is to identify the unknown white powder from the crime scene by matching it with one of the powders found on the clothes of one of the suspects.

Working in pairs, you will first determine the solubility of either Epsom salt, citric acid or sodium chloride. Add one level teaspoon of the solid to a beaker with 50 mL of water and stir with a glass rod or wooden stirrer for 1 – 2 minutes, or until the solid is dissolved. If all of the solid dissolves, add another level teaspoon and again stir. Continue adding 1 teaspoon and stirring until the water is saturated. Report the number of teaspoons that dissolved in the 50 mL of water in the table below. When you are done, take a vial of the unknown powder found at the crime scene and determine its solubility in water.

Observations:

Suspect	White Powder	Teaspoons Needed to Saturate 50 mL of Water
#1	Epsom salt	
#2	calcium carbonate	
#3	sodium chloride	
	Crime-scene powder	

Questions:

1) Why is it important that all the pairs of students use a level teaspoon when determining the solubility of their solid?

2) Based on the results of your tests, which suspect is most likely the thief? Explain.

3) Based on the class results, how much sodium chloride would it take to saturate 350 mL of water? Show how you came up with your answer.

Name: _____   **Score:** _____   **points**

1) Can the relative amounts of the components of a compound vary? Can the relative amounts of the components of a mixture vary? Explain. Include the following examples as part of your explanation: Water is a compound and green ink is a mixture.

2) Based on personal experience or knowledge, would you expect to be able to add more of a typical solid to cold water or to warm water before it becomes saturated? Describe the experience on which you are basing your answer. If you don't have any relevant personal experiences, try it out by adding salt or sugar to glasses of hot and cold water.

3) Is it easier to float in a fresh water lake or in the ocean? Explain your answer based on what you learned in *Part B*.

Continued on Back →

4) Imagine you have decided to transfer your saltwater fish from a 40-gallon tank to a 250-gallon tank.

   a) In your old tank you needed to add 15 cups of specially formulated salt to keep the fish happy. How much salt will you need for you new tank to have the same concentration of salt that you had in the small tank? Show your work.

   b) Saltwater fish need to have enough space. If you were able to keep two, 4-inch fish in your small tank, how many 4-inch fish could you keep in your new tank and still give them the same amount of room? Show your work.

5) In very simple terms, when a scientist examines a sample of DNA, he or she begins by isolating the DNA to be compared (for example, the DNA of the father and child in the case of a paternity suit). The DNA is then cut up into shorter fragments and sorted by size. DNA, being negatively charged, is pulled toward the positively charged electrode, with larger fragments moving more slowly than smaller ones. The resulting separations for the different DNA samples can then be compared. How is this process similar to the separation of ink that you performed in *Part A*? If you are interested, check out http://science.howstuffworks.com/dna-evidence.htm to find more information on how DNA testing works.

Activity 8b: Exploring Acids and Bases
COVER SHEET

Before Class
- Read the pages in your textbook dealing with "acids and bases." Your instructor may assign specific pages for you to read.
- Complete the Pre-Activity Problem Set on the next page.

Key Concepts
- Acids and bases are classes of compounds that exhibit characteristic behaviors when added to water.
- Acid/base indicators are compounds that turn various colors based on the pH of the solution to which they are added.

Learning Objectives
- You will be able to determine if a compound is likely to be an acid or a base.
- You will be able to describe in words and in drawings the behavior (including conductivity) of acidic and basic compounds when dissolved in water.
- You will be able to relate the different properties of strong and weak acids/bases to their behavior on a submicroscopic level.
- You will be able to calibrate an acid – base indicator and use the indicator to determine the pH of a substance.
- You will be able to rank the acidity of common household chemicals.
- You will be able to relate your understanding of acids and bases to practical situations.

Name: **Score:**   **points**

1) What is the definition of an acid?

2) What are some typical characteristics of acids?

3) Give an example of a common substance that is acidic. Give the name and formula of the acid that is present in the substance.

4) What is the definition of a base?

5) What are some typical properties of bases?

6) Acids and bases can react with each other to form water and a salt. An example of such an acid – base neutralization reaction occurs between aqueous solutions of HBr and KOH. What is the salt that is formed when HBr and KOH neutralize each other?

7) Explain the difference between what happens when a strong acid and a weak acid are added to water.

Continued on Back →

8) For any solution to be considered neutral, what must be the pH?

9) Two samples of rainwater are examined and are found to have pH's of 4.5 (Sample #1) and 5.8 (Sample #2). Which sample is more acidic? Explain.

10) What acid is formed when carbon dioxide reacts with water?

CAUTION: Wear goggles during this activity. Always be careful when working with acids and bases. Avoid contact with skin. Immediately inform your instructor of any spills.

Instructions:
- Begin with a clean, dry well-plate. Working in pairs, half-fill one of the wells with 1.0 M NaOH and half-fill a second well with 1.0 M HCl.
- Use a conductivity probe to test the electrical conductivity of the two solutions. Be sure to rinse the electrodes with distilled water between each test so that you do not cross-contaminate your solutions.
- Electrical conductivity is determined by observing the light bulb on the conductivity probe. The brighter the light, the greater the conductivity. If no light is observed, the solution does not conduct.
- When you are finished, dispose of the solutions according to your instructor's directions. Rinse all glassware and electrodes with distilled water. Dry all equipment.
- **Note #1**: The "(aq)" after the formula for the acid or base means that it has been added to water to make a solution.
- **Note #2**: The number in front of the "*M*" indicates the concentration of the solution. The larger the number, the higher the concentration. In general, solutions less than 1 – 2 M can be considered "dilute" and solutions greater than 5 – 6 M can be considered "concentrated."

Observations:

Name	Formula	Acid or Base?	Type of Bonding Present?	Does It Conduct?
Sodium hydroxide, 1.0 M	NaOH(aq)			
Hydrochloric acid, 1.0 M	HCl(aq)			

Questions:

1) Based on what we have learned so far, we *shouldn't* be surprised that aqueous solutions of bases conduct electricity. Explain why.

2) Based on what we have learned so far, we *should* be surprised that aqueous solutions of acids conduct electricity. Explain why.

3) Using what you learned in your textbook reading, what ions are responsible for the electrical conductivity in the HCl(aq)?

Continued on Back →

4) Using the drawing of a beaker of concentrated NaOH(aq) below as a guide, draw two beakers: one of a dilute solution of NaOH(aq) and one of a solution of HCl(aq) that has the same concentration as the first beaker of NaOH. Note: For clarity, the water molecules have not been drawn.

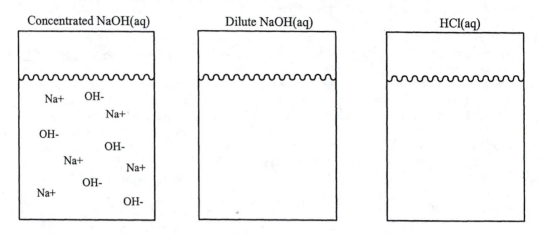

5) Which of the two beakers of NaOH in the previous question (concentrated or dilute) would you expect to exhibit greater conductivity? Explain.

Instructions:
- Begin with a clean, dry <u>well-plate</u>. Working in pairs, half-fill one of the wells with <u>1.0 M acetic acid</u> and half-fill a second well with <u>1.0 M HCl</u>.
- Use a <u>conductivity probe</u> to test the electrical conductivity of the two solutions. Be sure to rinse the electrodes between each test with <u>distilled water</u> so that you do not cross-contaminate your solutions. Look for any difference in how bright the lightbulbs are.
- When you are finished, dispose of the solutions according to your instructor's directions. Rinse all glassware and electrodes with distilled water. Dry all equipment.

Observations:

Name	Formula	Does It Conduct Strongly or Weakly?
Acetic acid, 1.0 M	$HC_2H_3O_2(aq)$	
Hydrochloric acid, 1.0 M	$HCl(aq)$	

Questions:
1) Based on your observations, which acid (acetic or hydrochloric) would you conclude is generating more ions when dissolved in water? Explain.

2) Draw pictures of $HCl(aq)$ and $HC_2H_3O_2(aq)$ that are consistent with your observations. Remember the two solutions have the same concentration (i.e., the same number of acid molecules were added to each beaker).

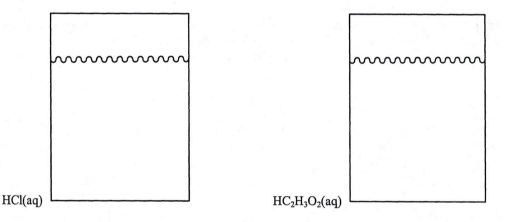

HCl(aq) $HC_2H_3O_2(aq)$

3) Chemists use the term *strong acid* to describe acids that are good at doing what acids do (that is, they are good at donating H^+ ions). By contrast, *weak acids* are not as good at donating H^+ ions. Which of these two acids (acetic or hydrochloric) would be described as *strong*? As *weak*? Explain.

Instructions:

- Begin with a clean, dry <u>well-plate</u>. Working in pairs, half-fill one of the wells with <u>1.0 M ammmonia</u> and half-fill a second well with <u>1.0 M NaOH</u>.
- Use a <u>conductivity probe</u> to test the electrical conductivity of the two solutions. Be sure to rinse the electrodes between each test with <u>distilled water</u> so that you do not cross-contaminate your solutions. Look for any difference in how bright the lightbulbs are.
- When you are finished, dispose of the solutions according to your instructor's directions. Rinse all glassware and electrodes with distilled water. Dry all equipment.

Observations:

Name	Formula	Does It Conduct Strongly or Weakly?
Ammonia, 1.0 M	NH_3 (aq)	
Sodium hydroxide, 1.0 M	NaOH (aq)	

1) Sodium hydroxide (NaOH) is a typical strong base. Would you classify NH_3 as a strong base or a weak base? Explain your answer based on your conductivity findings.

2) All the strong bases are ionic compounds made from a metal ion from group 1 (like Na^+ or K^+) or group 2 (like Ca^{2+} or Ba^{2+}) combined with an OH^- ion. Write the name and formula for the strong base made using Sr^{2+} ions.

3) Since the strong bases all have an OH^- in their formula, it is easy to see how they fit the definition of bases. It is not clear, however, from its formula how NH_3 can act as a base. Explain how NH_3 can generate OH^- ions in water and thereby fit the definition of a base.

4) Using the drawing of a breaker of NaOH(aq) below as a guide, draw two beakers—one of NH_3 and the other with $Ca(OH)_2$. All three beakers should have the same concentration as the beaker of NaOH.

NaOH(aq)

Na+   OH-

OH-   Na+

        OH-   Na+

Na+

        OH-

NH_3(aq)

$Ca(OH)_2$(aq)

Part D: pH of Household Chemicals Using Acid – Base Indicators

Instructions:

1) Add 10 drops (or a small scoop in the case of solids) of each of the <u>provided common household chemicals</u> to a separate clean, dry well of a <u>well-plate</u>. Add 5 drops of <u>cabbage juice</u>* to each of the chemicals. Record the color and pH in the table below. Discard the solutions according to your instructor's directions. Rinse/dry the well-plate.
2) Repeat step 1 but instead of cabbage juice, test each compound with <u>phenolphthalein</u>.
3) Repeat step 1 but instead of cabbage juice, test each compound with <u>phenol red</u>.
4) If available, record the pH of each of the household chemicals using the <u>pH meter</u>. Your instructor will show you how to use a pH meter to record the pH of each of the solutions. Be sure to rinse the pH meter with distilled water between each measurement.

Chemical	Cabbage Juice: Color	Cabbage Juice: pH **	Phenolphthalein: Color	Phenol Red: Color	pH Meter: pH

**Red cabbage juice turns the following colors at the pH values listed below:

pH	1 – 2	3 – 6	7	8	9 – 12	13
Color	Red	Violet	Blue	Green-blue	Green	Yellow-green

Question:

1) Rank the above household chemicals from most basic to most acidic using cabbage juice and the pH meter. Are the two rankings in agreement?

                          Most Basic   →   →   →   **Most Acidic**

According to
Cabbage Juice:

According to
pH Meter:

2) What color does phenolphthalein turn in acidic conditions? In basic conditions?

3) What color does phenol red turn in acidic conditions? In basic conditions?

* To make your own cabbage juice, cut a red cabbage into 1-inch cubes. Place the cabbage cubes in a blender and add enough water to cover the cabbage. Blend the mixture until the cabbage has been chopped into uniformly tiny pieces. Using a strainer, strain the liquid from the mixture into a holding container, pressing the cabbage to speed the straining. The strained liquid is the cabbage juice indicator. Keep refrigerated until ready to use.

Instructions #1: You will be recording the color and estimating the pH at each step. Pour about 100 mL of water into a 250-mL beaker. Add about 30 drops of cabbage juice indicator (should be enough to give a strong color). Stir with a spoon or stirring rod and record the color. Add about 10 drops of 1.0 M HCl to the solution. Stir and record the color. Add about 15 drops of 1.0 M NaOH. Stir and record the color. Add a small chip (about 1-cm cube) of dry ice (solid CO_2) to the solution. Record any color changes.

Observations:

Step	What Was Done?	Color	pH	Acidic or Basic?
#1	Water + cabbage juice			
#2	+ 10 drops HCl			
#3	+ 15 drops of NaOH			
#4	+ chip of dry ice			

Questions:
1) Why does the color after completing Step #2 make sense?

2) What two chemicals are formed when HCl and NaOH neutralize each other?

3) Why does the color of the solution after completing Step #4 make sense?

Instructions #2: Half-fill a small plastic cup with water. Add 3 drops of 0.10 M $NaHCO_3$(aq) to the cup along with 3 drops of phenol red. Using a clean straw that you have just unwrapped, carefully blow bubbles into the solution until you see a color change. Record your observations. Refer to *Part D* for the colors of phenol red in acids and bases. **CAUTION:** You must wear safety goggles during this activity.

Observations:
Color Before Blowing Bubbles: Color After Blowing Bubbles:

Questions:
1) Is $NaHCO_3$ an acid or a base? Explain.

2) Does your breath make the solution acidic or basic? Explain.

3) What component in your breath is responsible for the answer in Question #2? Hint: Look back at the Pre-Activity Problem Set.

Name: **Score:**   **points**

1) Find the name and the formula of the acid used by some ants as a weapon.
    Name of Acid: Formula of Acid:

    Source of Info:

2) Two students are asked to each make up 1.0 L of an acidic solution. The only requirement is that each solution must have the same pH. The first student chooses to make her solution by adding a weak acid (for example, HCN) to the water. The second student makes his solution by adding a strong acid (for example, HCl) to the water. Both students end up correctly making their solution with the same pH. Explain how this is possible.

3) Draw pictures corresponding to the two solutions from Question #2 above.

Solution Made with Weak Acid (HCN):   Solution Made with Strong Acid (HCl):

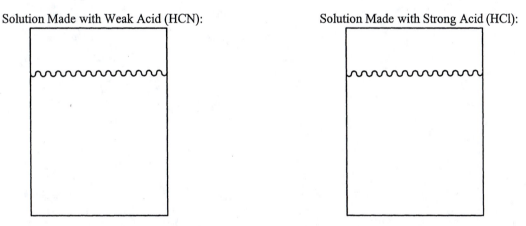

4) A sample of garden soil is found to have a pH of 5.0. The plants you wish to grow require a pH of 7.5. Are there too many hydrogen ions in the soil or too few? Would you need to add an acid or a base to the soil to correct the pH? Explain your answers.

Activity 8c: Determination of the Sugar in Soda
COVER SHEET

Before Class
- Read the pages in your textbook dealing with "density." Your instructor may assign specific pages for you to read.
- Complete the Pre-Activity Problem Set on the next page. You will need to use the Internet for some parts of the Pre-Activity Problem Set.

Key Concepts
- The density of a sugar – water solution is related to the concentration of the sugar.
- The density of a sample of soda depends primarily on the amount of sugar it contains.
- The equation for the best-fit line on a calibration curve can be used to determine unknown concentrations.
- Soda has a lot of sugar in it.

Learning Objectives
- You will be able to determine the density of a liquid.
- You will be able to design your own experimental protocol.
- You will be able to construct and interpret calibration curves.

Name: Score: points

Background: Because lead ions (Pb^{2+}) have been linked to anemia, kidney damage, and problems related to mental development in children, it is important to know the concentration of lead in municipal drinking water. To determine the amount of Pb^{2+} in a sample of water, a scientist can make what is called a "calibration curve." Calibration curves establish a linear relationship between the concentration of the chemical of interest (here, the Pb^{2+}) and some measurable property of the solution. For example, it is known that the amount of light that a sample of water absorbs has a linear relationship to the amount of lead in the sample. To make the calibration curve, the scientist will need to measure the absorbance of a series of "standard solutions" made with known concentrations of Pb^{2+}, as shown in the following instructions.

Instructions: Use a blank sheet of graph paper from the end of this manual and the data in the table below to make a calibration curve relating the concentration of Pb^{2+} ions in solution to the amount of light absorbed by the sample. Note that the "concentration of lead" is the *independent variable* (the one the scientist controls) and is plotted on the x-axis. The dependent variable (in this case, the "absorbance of sample") is plotted on the y-axis.

Standard	Concentration of Pb^{2+} ions (µg/mL of Water)	Absorbance of Sample
#1	0.0	0.00
#2	0.5	0.15
#3	1.0	0.30
#4	1.5	0.42
#5	2.0	0.53
#6	2.5	0.69
#7	3.0	0.80

Questions:

1) Attach your calibration curve to your Pre-Activity Problem Set. Make sure your graph has labeled axes, a title, and a best-fit straight line drawn using the data points (do NOT "connect the dots").

2) The equation of the best-fit line can be found using $y = mx + b$. Where m is the slope of the line and b indicates where the line crosses the y-axis. The slope of the line can be determined by calculating the rise/run (i.e., $\Delta y/\Delta x$). What is the equation of the best-fit line? Show all your work.

3) If the local community has requirements that the concentration of lead must be below 1.3 µg/mL of water, is the drinking water safe if it is found to have an absorbance of 0.32? Defend your answer using the equation of the best-fit line from your calibration curve.

Continued on Back →

In this activity, you will be hired by your local school board to help determine if it should ban the sale of soda at school. To prepare yourself, search the Internet to find two or three news stories related to banning soda in schools.

4) List two URLs for Web news articles dealing with banning soda in schools.

5) Briefly summarize the pros and cons related to banning soda in schools.

Instructions: Working in small groups, design an experimental protocol that you can use to determine the amount of sugar in a <u>20-oz bottle of soda</u>. Use the space below and on the back to record your procedure. Some key questions your group should discuss include:

- What <u>equipment</u> will you need?
- What will you plot on your calibration curve?
- What range will you use for the dependent variable?
- How many data points (different standard solutions) will your calibration curve have?
- How will the class divide the work of making the calibration curve (i.e., how many data points will each group of students make?)
- Should there be multiple trials for each standard solution?
- How can you ensure that each pair of students makes the standard solutions the same way? For example, how do you make sure everyone measures the same amount of sugar and how much water should be used when making each data point?

After each group comes up with a plan, the class will discuss the different possibilities and agree on a procedure to be carried out by the whole class working in pairs.

Instructions: Working in your small groups, carry out the experiment described in *Part A*. Use the space below and on the back to outline exactly what your group did and to record all of the class data.

Name: _____   **Score:** _____ **points**

| **Some Helpful Information:** | 1 tsp of sugar weighs about 4 g and has about 15 Calories |
| | A 20-oz bottle of soda has a volume of 600 mL |

1) Attach the page from *Part B* (outlining your experiment and the class data) to your Post-Activity Problem Set. Also, attach a copy (either computer-generated or neatly hand-drawn) of the calibration curve you made with the class data. Make sure your graph takes up the full page, has a title, a labeled axis, and a best-fit line.

2) What is the equation for the best-fit line for your calibration curve? Show all your work if the best-fit line was not determined by computer.

3) What was the volume of your soda sample?

4) What was the measured density of your soda sample? Show your work.

5) Use the equation for the best-fit line from your calibration curve to calculate the number of grams of sugar in your soda sample. Show your work.

6) Calculate the number of grams of sugar in a 20-oz bottle of soda. Show all your work.

Continued on Back →

7) If the Recommended Daily Intake (RDI) for added sugar is 40 g, what percent of the RDI is found in a 20-oz bottle of soda? Show your work.

8) Using your results, calculate the number of Calories due to sugar in a 20-oz bottle of soda. Show your work.

9) In your opinion, should the sale of soda be banned in schools? Briefly explain.

Activity 8d: Making Water Safe to Drink
COVER SHEET

Before Class
- Read the pages in your textbook dealing with "drinking water" and "wastewater treatment." Your instructor may assign specific pages for you to read.
- Complete the Pre-Activity Problem Set on the next page.

Key Concepts
- We rarely encounter water in a pure state. Most of the water around us, including rainfall, bottled and tap water, and the water in rivers and lakes, are all mixtures.
- Some things that make lake and river water healthy do not make for good drinking water.
- Chemistry is applied on many levels to making water safe to drink.

Learning Objectives
- You will be able to explain the various stages of a water treatment process.
- You will be able to research, on your own, a topic related to water treatment.

Pre-Activity Problem Set: Activity 8d

Name: _____ Score: _____ points

1) The following questions refer to mixtures made up with water.
   a) Give an example where water is part of a heterogeneous mixture.

   b) Give an example where water is part of a homogeneous mixture (a solution). What is the solute and what is the solvent?

   c) Discuss how the way that hydrogen and oxygen atoms combine to make a water molecule is different from the way a solvent and a solute combine to make a solution.

2) Following the two examples, complete the table below by listing four more things that might be found in a sample of water. What is the source of each of these things? Is each thing typical of healthy lake water? Would each thing be acceptable in good drinking water?

Thing	Possible Source	Good for Lake?	Good to Drink?
Lead	Enters water from old plumbing	No	No
Fish	Naturally occurring	Yes	No

3) Is your drinking water primarily municipal, bottled, or from a well? Do you use any additional water treatment (for example, water softening or a filtering system) on your drinking water? Why?

Continued on Back →

4) After putting each of the steps in the typical water treatment process in the correct order, match each of the steps with its description. Use the clues in the descriptions to help you order the steps.

Step	Order #		Description
Filtration	_____	•	Small amounts of a chemical are added to kill bacteria or other microorganisms that may be in the water.
Sedimentation	_____	•	Water is placed in a reservoir where disinfection can take place. The water then flows to homes.
Coagulation	_____	•	A clumping agent is added to the water. The clumping agent forms sticky particles called floc that attract suspended dirt.
Storage	_____	•	Water passes through sand, gravel, and charcoal filters that help remove even smaller particles.
Disinfection	_____	•	The heavy particles (floc) settle out and the clear water moves on to filtration.

Instructions #1: Your instructor has prepared <u>two beakers with samples of "swamp water"</u> (made by mixing about 1 cup of dirt and clay in a milk jug full of water). The beaker labeled "treated" had about 2 tsp of alum added and was stirred for 10 minutes, while the sample labeled "untreated" was left alone. Both samples were allowed to sit overnight. Without disturbing the solutions, carefully look at both samples and record your observations.

Observations:
Treated Sample:

Untreated Sample:

Questions:
1) Which one of the two samples is cleaner? Which one do you think would be easier to filter? Explain.

2) To which two parts of the municipal water treatment cycle described in the Pre-Activity Problem Set (Question #4) does the treatment with alum correspond?

Continued on Back →

Instructions #2: If one is not already prepared, cut a <u>small water bottle</u> in half with a pair of <u>scissors</u>. The top part of the bottle will be inverted and used to treat our water, and the bottom of the bottle will be our collection container. Stuff a <u>cotton ball</u> in the mouth of the water bottle. Invert the top half of the bottle so that is it stuck, upside-down, in the bottom half of the bottle. Make your water treatment column by filling your upside-down bottle with various samples of <u>gravel, sand, rocks, cotton balls, and coffee filters</u>. You can choose to test just one material or you can layer as many of the samples as you want. Take a <u>cup filled with "swamp water"</u> and pour half of the water into your water treatment column. Compare the treated water that comes out the bottom to the untreated swamp water. Compare your treated water with the rest of the class. Record your observations. Dispose of the treated water and the contents of your water treatment column according to your instructor's directions.

Observations:

Questions:

1) Is your water treatment column a homogenous or heterogeneous mixture? How about the pretreated swamp water? The treated swamp water?

2) Who in the class had the cleanest treated water? What materials did they use in their water treatment column and in what order?

3) Are there any other materials that you can think of that might make a good water treatment column? What is it about those materials that you think would make them well suited for the job?

4) To which step of the municipal water treatment cycle described in the Pre-Activity Problem Set (Question #4) does your water treatment column correspond?

Explorations in Conceptual Chemistry: Activity 8d

Instructions#1: Your instructor has prepared a sample of "smelly water" (made by adding a few tablespoons of ammonia to a milk jug full of water). Get a <u>small beaker</u> half-filled with "<u>smelly water</u>" and a <u>second small beaker</u> half-filled with <u>baking soda</u>. **Carefully** waft the "smelly water" toward your nose and record your observations. Add a level <u>teaspoon</u> of <u>baking soda</u> to the water and stir for about 1 minute. Again, **carefully** smell the water. Continue adding single teaspoons of baking soda and stir until the smell is gone. Record how many teaspoons of baking soda you needed to add. Dispose of the treated water according to you instructor's directions.

Observations:

Question:

1) During municipal water processing, odors are often treated by aerating the water. Explain why you think cascading the water through a column of air can remove unpleasant smelling gases like sulfur compounds as well as radioactive gaseous radon.

Instructions #2: Your instructor has prepared a sample of "contaminated water" (made by adding a few drops of food coloring to a milk jug full of water). Get <u>two small beakers</u>, each half-filled with "<u>contaminated water</u>," and a <u>third small beaker</u> half-filled with <u>bleach</u>. Record the color of the water. Add a teaspoon of bleach and stir for about 1 minute. Continue adding single teaspoons of bleach and stirring until the color is gone. Record how many teaspoons of bleach you needed to add. Dispose of the treated water according to your instructor's directions.

Observations:

Questions:

1) The bleach works really well at breaking down the molecules of the food coloring (which is why it is added to laundry to whiten clothes), but there is an obvious problem with just dumping in a ton of chemicals such as bleach to treat drinking water. Explain.

2) Rather than bleach, water treatment plants use chlorine or ozone to kill bacteria or microorganisms. To which step of the municipal water treatment cycle described in the Pre-Activity Problem Set (Question #4) does your treatment with chemicals correspond?

Post-Activity Problem Set: Activity 8d

Name: _____ **Score:** _____ **points**

Select **two of the four** water treatment topics listed below to research. Answer each of the questions listed for your chosen topics. Use the back of this page as necessary. If you type up your answers, attach them to this sheet. Indicate all the sources of your information.

1) Trihalomethanes in Water
   - What are trihalomethanes and how do they form in drinking water?
   - Why can't we just do away with the conditions that lead to their formation?
   - What health risks do they pose in drinking water?

2) Hard Water
   - What is hard water and what are some problems associated with hard water?
   - How is hard water treated and are there any drawbacks to the treatment process?

3) Reverse Osmosis
   - What is reverse osmosis and how is it used to treat water?
   - Are there any drawbacks to treating water by reverse osmosis? Explain.

4) Fluoride in Water
   - Why is fluoride added to water and how does fluoride work?
   - What is the controversy behind adding fluoride to water?

Unit 9: Important Aspects of Chemical Reactions

New learners of chemistry often have difficulty appreciating the submicroscopic and macroscopic significance contained within the symbolic representation known as a chemical equation. Many students also struggle with the process of balancing chemical equations as is required by the law of conservation of mass. Unit 9 is designed to help you get beyond the symbols and arrows so you can gain a true conceptual understanding for what a chemical reaction is telling us.

Unit 9 also provides us with the opportunity to tie heat and particle motion in with chemical reactions. In addition to an introduction to heat changes in chemical reactions, you will also experimentally determine the amount of heat, in the form of Calories, found in a sample of peanuts.

We finish Unit 9 and our tour of the world of chemistry by exploring the factors that determine the rate at which a reaction occurs and with an introduction to electrochemistry. Electrochemical processes are crucial to biological life (they enable our heart to beat and our brain to think) as well as to many of the comforts that we take for granted (such as batteries for cell phones, pacemakers, and automobiles).

Activity 9a: An Introduction to Chemical Reactions
COVER SHEET

Before Class

- Read the pages in your textbook dealing with "chemical reactions." Your instructor may assign specific pages for you to read.
- Complete the Pre-Activity Problem Set on the next page.

Key Concepts

- During physical and chemical changes, atoms are neither created nor destroyed. This principal is known as the law of conservation of mass.
- Before and after a physical change, the same compounds are present.
- During a chemical change/reaction, bonds between atoms are broken and new bonds are made. As a result, different compounds are present before and after a chemical change.
- The changes that occur during a chemical change can be summarized by a balanced chemical equation.
- Some elements typically occur in nature as diatomic molecules (i.e., H_2, N_2, O_2, F_2, Cl_2, Br_2, and I_2).
- Carbon dioxide and water are produced when any compound containing just C and H is burned in oxygen.

Learning Objectives

- You will be able to write chemical reactions when given written descriptions, and write written descriptions from chemical reactions.
- You will be able to balance chemical reactions.
- You will be able to convey, in words and in drawings, what is occurring during chemical reactions.

Pre–Activity Problem Set: Activity 9a

Name: _____ **Score:** _____ **points**

1) What is the difference between a physical change and a chemical change/reaction?

2) Identify (circle your answer) each of the following as either a physical or chemical change:

   a) A tree growing   Physical change   Chemical change

   b) Chopping down a tree   Physical change   Chemical change

   c) Ripping a piece of paper   Physical change   Chemical change

   d) Burning a piece of paper   Physical change   Chemical change

3) The written description and symbolic chemical reaction below describe the same chemical change. Compare the written and symbolic representations and fill in the table that follows.

   Written Description: Solid sodium cyanide reacts with an aqueous solution of hydrochloric acid to produce an aqueous solution of sodium chloride and bubbles of hydrogen cyanide gas.

   Symbolic Chemical Reaction: $NaCN(s) + HCl(aq) \rightarrow NaCl(aq) + HCN(g)$

   Table:

Symbol	What It Represents
NaCN	
(s)	
+	
HCl	
(aq)	
→	
NaCl	
(aq)	
+	
HCN	
(g)	

4) Draw a circle around each of the reactants and a box around each of the products in the previous question.

5) What is the difference between NaCl(l) and NaCl(aq)? As part of your answer, explain how each could be made starting with NaCl(s).

Continued on Back →

6) Translate the following symbolic chemical reaction into a written description:

Reaction #1: $NaCl(aq) + AgNO_3(aq) \rightarrow AgCl(s) + NaNO_3(aq)$

7) You may know that vinegar (aqueous acetic acid) and baking soda (sodium bicarbonate) react to form a gas that can be used to fill a balloon or launch a small toy boat or rocket. Translate the following written description into a symbolic chemical reaction:

Reaction #2: Aqueous acetic acid ($HC_2H_3O_2$) reacts with solid sodium bicarbonate to produce water, an aqueous solution of sodium acetate, and carbon dioxide gas.

8) In today's activity, we examine the hows and whys of "balancing" a symbolic chemical reaction. In preparation, imagine you need to make the shape on the right side of the arrow (*Object C*) using only the two shapes on the left side of the arrow (*Objects A* and *B*). If needed, you are allowed to cut up *Objects A* and *B* along their dividing lines to make two individual white or gray square boxes. What is the smallest whole-number ratio that you'll need of *Objects A* and *B* to make a whole number of *Object C* with no pieces of *Objects A* or *B* left over? Show you answer pictorially in the empty space below.

Hint: In the current drawing (one each of *Objects A* and *B*), you would have one extra white square and would be missing one grey square when you tried to make *Object C*. If, for example, you had one of *Object A* and two of *Object B*, you could make one of *Object C*, but you would have one white square and one gray square left over.

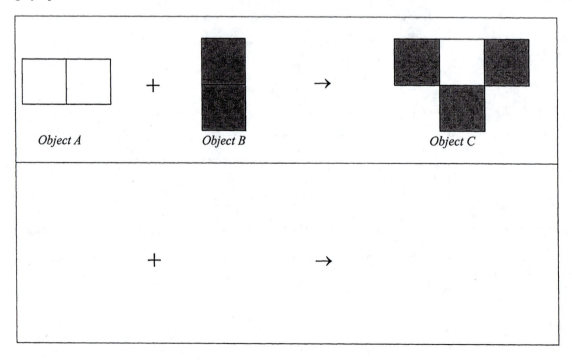

CAUTION: Physical contact with silver nitrate will stain exposed skin black for about 1 week!!! Wear your goggles during this activity. All waste goes in the labeled waste container.

Instructions: Use <u>well-plates</u> and 5 – 10 drops of each chemical (use only a pinch of baking soda) to perform Reaction #1 and Reaction #2 (from Pre-Activity Problem Set Questions #6 and #7). As you carry out Reactions #1 and #2, record your observations in the spaces below.

Reaction #1: $\underline{NaCl(aq)} + \underline{AgNO_3(aq)} \rightarrow AgCl(s) + NaNO_3(aq)$

Reaction #2: <u>Aqueous acetic acid</u> ($HC_2H_3O_2$) reacts with <u>solid sodium bicarbonate</u> (baking soda) to produce water, an aqueous solution of sodium acetate, and carbon dioxide gas.

Observations:
Reaction #1:

Reaction #2:

Questions:

1) How do your macroscopic observations for Reactions #1 and #2 relate to what the symbolic chemical reactions indicate is happening on the submicroscopic level?

2) Fill in the boxes below with drawings that indicate what is present in each of the corresponding beakers during <u>Reaction #1</u>. NOTE: For clarity, water molecules have not been shown.

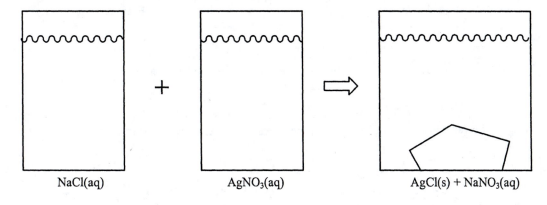

NaCl(aq)   AgNO₃(aq)   AgCl(s) + NaNO₃(aq)

Case #1: The unbalanced chemical equation for the reaction of methane gas and oxygen gas to produce gaseous carbon dioxide and water vapor

1) Draw Lewis dot structures and VSEPR shapes (with labeled angles) for each of the molecules in the unbalanced reaction below. Use the model kits to build the molecules if you are having trouble figuring out the VSEPR shapes.

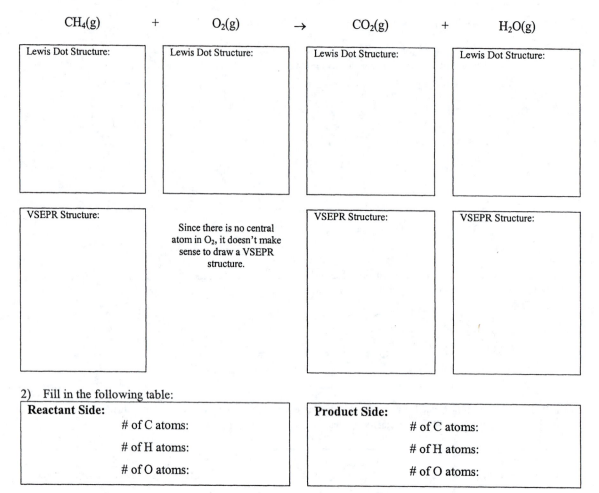

$$CH_4(g) \quad + \quad O_2(g) \quad \rightarrow \quad CO_2(g) \quad + \quad H_2O(g)$$

Lewis Dot Structure:

Lewis Dot Structure:

Lewis Dot Structure:

Lewis Dot Structure:

VSEPR Structure:

Since there is no central atom in O_2, it doesn't make sense to draw a VSEPR structure.

VSEPR Structure:

VSEPR Structure:

2) Fill in the following table:

Reactant Side:	Product Side:
# of C atoms:	# of C atoms:
# of H atoms:	# of H atoms:
# of O atoms:	# of O atoms:

3) Which atoms (and how many of them) in the above reaction seem to have disappeared?

4) Which atoms (and how many of them) in the above reaction seem to have appeared from nowhere?

5) What chemical law does this "disappearing" and "appearing" of atoms disobey?

Continued on Next Page →

Case #2: The rewritten chemical equation for the reaction of methane gas and oxygen gas to produce gaseous carbon dioxide and water vapor

$$CH_4(g) \quad + \quad 2\ O_2(g) \quad \rightarrow \quad CO_2(g) \quad + \quad 2\ H_2O(g)$$

6) Again, draw the Lewis dot structures for each of the molecules in the above chemical reaction. This time, however, draw the appropriate number of molecules as indicated by the coefficients.

Lewis Dot Structure:	Lewis Dot Structure:	Lewis Dot Structure:	Lewis Dot Structure:

7) Using your drawings from Question #6, fill in the following table:

Reactant Side:	Product Side:
# of C atoms:	# of C atoms:
# of H atoms:	# of H atoms:
# of O atoms:	# of O atoms:

8) Is the chemical equation in **Case #2** balanced? Explain.

9) Going from reactants to products, how many of each type of bond (for example, C—H bond) were broken? How many of each type of new bond were formed? Which of the Key Concepts found on the COVER SHEET of this activity ties in with this observation?

- Once a reaction is balanced, it can be "read" as if individual molecules (in the correct ratio) are undergoing the chemical reaction.

$CH_4(g)$	+	$2\ O_2(g)$	$\rightarrow$	$CO_2(g)$	+	$2\ H_2O(g)$
1 molecule of gaseous methane	reacts with	2 molecules of oxygen gas	to produce	1 molecule of gaseous carbon dioxide	and	2 molecules of water vapor

Of course, chemists almost never work with single molecules. Luckily, however, the numbers can be scaled up as long as the ratio is kept the same.

1) With this in mind, how many molecules of water can be made, according to the above reaction, if you react 1.5 billion molecules of CH_4 (assuming there is plenty of O_2)?

2) How many molecules of oxygen are needed to make 30 million molecules of carbon dioxide, according to the above reaction?

- The conservation of mass during a chemical reaction can be verified by comparing the masses of the individual reactant and product molecules.

3) What is the total mass (in amu) of the reactants?

Mass of 1 molecule of methane =	amu
Mass of 1 molecule of oxygen =	amu
Mass of 1 molecule of oxygen =	amu
Total mass =	amu

4) What is the total mass (in amu) of the products?

Mass of 1 molecule of carbon dioxide =	amu
Mass of 1 molecule of water =	amu
Mass of 1 molecule of water =	amu
Total mass =	amu

5) What is significant about the results of Questions #3 and #4? Would this be true of an unbalanced reaction? Explain.

- NEVER balance a chemical reaction by changing the subscripts.

6) The following reaction is also "balanced:" $CH_4(g) + O_3(g) \rightarrow CO_2(g) + H_4O(g)$
   Why is it incorrect to balance the equation in **Case #1** (back in *Part B*) in this fashion?

Part D: Practice Balancing Chemical Reactions

Instructions: Balancing reactions requires practice, and sometimes trial and error. Begin by balancing elements that only occur in one place on each side of the equation. Typically, this means leaving oxygen and/or hydrogen to balance last. The following questions will give you some practice.

NOTE: When writing chemical reactions, certain elements are diatomic (H_2, N_2, O_2, F_2, Cl_2, I_2 and Br_2). For example, "hydrogen gas" would be translated as "$H_2(g)$." All other pure elements are simply written as a single atom. For example, "solid metallic iron" would translate as "$Fe(s)$."

Questions:

1) When an aqueous solution of copper (II) nitrate is mixed with aqueous sodium carbonate, the result is the formation of solid copper (II) carbonate and aqueous sodium nitrate.

   a) Write the balanced chemical reaction for the written description above.

   b) Fill in the boxes below with drawings that indicate what is present in each of the corresponding beakers. NOTE: For clarity, water molecules have not been shown.

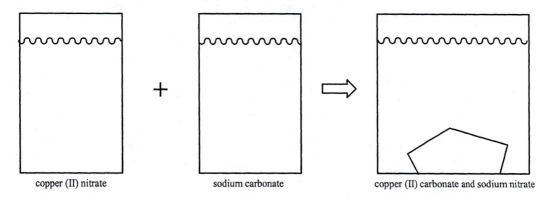

copper (II) nitrate	sodium carbonate	copper (II) carbonate and sodium nitrate

   c) How does your drawing symbolically indicate what you would expect to observe on a macroscopic level if you actually carried out the reaction?

2) In solid rocket boosters for spacecrafts, solid ammonium perchlorate and solid aluminum metal react to form solid aluminum oxide, solid aluminum chloride, gaseous nitrogen monoxide, and water vapor. Write the balanced chemical reaction for this process.

Unbalanced Reactions   **Answers**

a) ___ $C_3H_8(g)$ + ___ $O_2(g)$ → ___ $CO_2(g)$ + ___ $H_2O(g)$ | 1, 5, 3, 4 |

b) ___ $Al(NO_3)_3(aq)$ + ___ $Na_2CO_3(aq)$ → ___ $NaNO_3(aq)$ + ___ $Al_2(CO_3)_3(s)$ | 2, 3, 6, 1 |

c) ___ $FeS_2(s)$ + ___ $O_2(g)$ → ___ $Fe_2O_3(s)$ + ___ $SO_2(g)$ | 4, 11, 2, 8 |

d) ___ $Ba(s)$ + ___ $H_2O(l)$ → ___ $Ba(OH)_2(aq)$ + ___ $H_2(g)$ | 1, 2, 1, 1 |

e) ___ $Cl_2(g)$ + ___ $C_2H_6(g)$ → ___ $C_2HCl_5(g)$ + ___ $HCl(g)$ | 5, 1, 1, 5 |

f) ___ $NH_3(g)$ + ___ $O_2(g)$ → ___ $NO(g)$ + ___ $H_2O(g)$ | 4, 5, 4, 6 |

g) ___ $Cu(s)$ + ___ $HNO_3(aq)$ → ___ $Cu(NO_3)_2(aq)$ + ___ $H_2O(l)$ + ___ $NO_2(g)$ | 1, 4, 1, 2, 2 |

h) ___ $PbS(s)$ + ___ $H_2O_2(aq)$ → ___ $PbSO_4(s)$ + ___ $H_2O(l)$ | 1, 4, 1, 4 |

Name: _____ **Score:** _____ **points**

1) Ammonia (NH_3) is an important chemical used mainly as a fertilizer. Gaseous ammonia is synthesized (made) by reacting hydrogen gas and nitrogen gas under high temperature and pressure.

   a) Write the balanced chemical reaction for the synthesis of ammonia.

   b) Use Lewis dot structures to fill in the boxes below with the correct number of each reactant and product molecule in the ratio indicated by the balanced reaction for the synthesis of ammonia.

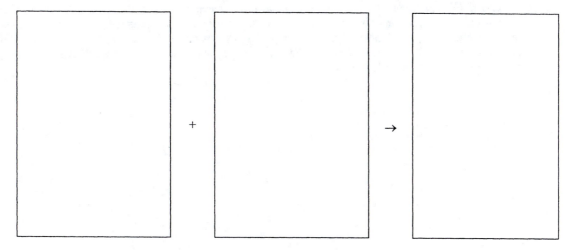

   c) Explain how Question #8 from the Pre-Activity Problem Set could be used to help illustrate how to balance the chemical reaction for the synthesis of ammonia.

Continued on Back →

2) Write the balanced chemical reaction that occurs in a car battery where solid metallic lead and solid lead(IV) oxide react with aqueous sulfuric acid (H_2SO_4) to produce solid lead(II) sulfate and liquid water.

3) Nitric acid (HNO_3) is a component of acid rain that forms when rainwater reacts with certain nitrogen-containing emissions from car exhaust and power plants. In addition to lowering the pH of soil and lakes to a degree that can kill plants and fish, acid rain also destroys statues and buildings by converting marble (calcium carbonate) into aqueous calcium nitrate, gaseous carbon dioxide, and water. Write the balanced chemical reaction between nitric acid and marble.

4) In our bodies, carbohydrates and other sugars can be broken down or converted to glucose ($C_6H_{12}O_6$) and used as fuel. In any combustion process, such as human metabolism, a fuel containing carbon and hydrogen (in this case, glucose) will react with oxygen to produce carbon dioxide and water vapor. Write the balanced chemical reaction for the combustion of glucose.

Activity 9b: Endothermic and Exothermic Chemical Reactions
COVER SHEET

Before Class
- Read the pages in your textbook dealing with "endothermic and exothermic reactions." Your instructor may assign specific pages for you to read.
- Complete the Pre-Activity Problem Set on the next page.

Key Concepts
- Chemical reactions require the breaking and forming of chemical bonds. This rearrangement of atoms to form new compounds involves changes in energy.
- Some chemical reactions release energy while others absorb energy.

Learning Objectives
- You will be able to define basic terms related to heat changes during chemical reactions.
- You will be able to draw potential-energy diagrams for endothermic and exothermic reactions.
- You will be able to explain observations in terms of the heat changes that occur during chemical reactions.

Pre-Activity Problem Set: Activity 9b

Name: _____   Score: _____ points

1) The following questions refer to exothermic reactions.
   a) Define the term *exothermic*.

   b) Would you expect a beaker that contains an exothermic reaction to feel hot or cold to the touch? Explain.

   c) Draw an example of a generic potential-energy diagram for an exothermic reaction.

2) The following questions refer to endothermic reactions.
   a) Define the term *endothermic*.

   b) Would you expect a beaker that contains an endothermic reaction to feel hot or cold to the touch? Explain.

   c) Draw an example of a generic potential-energy diagram for an endothermic reaction.

Continued on Back →

3) We are already familiar with physical changes that absorb or release heat.
   a) List all of the endothermic phase changes.

   b) List all of the exothermic phase changes.

4) Is the chemical change occurring within an egg when it cooks endothermic or exothermic? Explain.

5) Define the term *bond energy*.

6) Would you expect it to take more energy to break a C—C bond or a C=C bond? Explain your answer.

Instructions #1: Put a heaping <u>teaspoon</u> of <u>solid CaCl$_2$</u> in the corner of a <u>resealable plastic bag</u>. Twist the corner of the bag so that the CaCl$_2$ is closed off from the rest of the bag. Add a teaspoon of <u>water</u> to the baggie (don't allow the water and calcium chloride to mix yet). Seal the bag. Allow the calcium chloride to mix with the water while you hold the bag in your hand. Record your observations. Dispose of the materials according to your instructor's directions.

Observations:

Questions:
1) What is the name of CaCl$_2$?

2) Does the process of dissolving CaCl$_2$ in water require an input or a release of energy? Explain your answer.

3) Could the process in Question #2 be used in a hot pack (to put on aching muscles) or a cold pack (to keep a can of soda cold)? Briefly explain.

Instructions #2: Repeat Instructions #1 above using a <u>new plastic bag</u> and <u>solid NH$_4$NO$_3$</u> instead of CaCl$_2$. Again record your observations. Dispose of the solution according to your instructor's directions.

Observations:

Questions:
1) What is the name of NH$_4$NO$_3$?

2) Does the process of dissolving NH$_4$NO$_3$ in water require an input or a release of energy? Explain your answer.

3) Could the process in Question #2 be used in a hot pack (to put on aching muscles) or a cold pack (to keep a can of soda cold)? Briefly explain.

Instructions: Wrap a <u>thermometer</u> in about one-third of a <u>steel wool pad</u> (the kind without soap). Place the thermometer and the steel wool in the <u>jar</u> and close the <u>lid</u>. Wait 2 – 3 minutes, then open the lid and record the temperature. Soak the steel wool in a <u>small beaker</u> with about <u>one-half inch of vinegar</u> for 1 minute. Squeeze the excess vinegar out of the steel wool. Again, wrap the wool around the thermometer and place the wool/thermometer in the jar, sealing the lid. Wait 5 minutes, then open the lid and record the temperature along with any other observations.

Observations:

	Temperature
Dry Steel Wool	
Steel Wool Soaked in Vinegar	

Other Notes:

Questions:

1) What is the common name given to the chemical reaction that is happening to the steel wool?

2) Why is the reaction in Question #1 happening faster than it might normally happen?

3) Is the reaction in Question #1 absorbing or releasing energy? Explain.

4) Where and in what form does the energy in Question #3 start out? Where and in what form does the energy end up?

5) Draw a potential-energy diagram for this process, showing "steel wool and vinegar" as the reactants and your answer to Question #1 as the product. Your diagram should clearly indicate whether the reaction is endothermic or exothermic.

Instructions: Your instructor will light a <u>candle</u> and drip some wax onto the middle of a <u>pie plate</u>. The wax will be used to fix the burning candle to the pie plate. After filling the bottom of the pie plate with <u>water</u> and a few drops of <u>food coloring</u> (so you can see it better), your instructor will put an upside-down <u>flask</u> over the burning candle so the open end of the flask is in the water. Record your observations (in both pictures and words).

Observations:

Questions:

1) Though the paraffin wax used to make candles isn't just one single compound, it can be approximated by the formula $C_{20}H_{42}$.

   a) Write the balanced chemical reaction for the combustion of paraffin.

   b) When watching a candle burn, it looks as though matter is not conserved (the candle seems to go away). How does the balanced chemical reaction help explain that matter is in fact conserved? Hint: Look at all of the states present in the balanced reaction.

   c) Is combustion an exothermic or endothermic reaction? How do you know? Give an example from your personal experience.

Continued on Back →

d) Using words and labeled drawings, fully explain your observations during "The Burning Candle Challenge." Hint: What does the combustion of paraffin do to the density of the air in the area around the candle flame?

Name: _____   Score: _____   points

1) A beaker containing an acid – base neutralization reaction will feel hot to the touch.
   a) Write the balanced reaction for the neutralization between HCl(aq) and KOH(aq).

   b) Draw a potential-energy diagram for the reaction between HCl(aq) and KOH(aq). Clearly label the reactants and products. Your diagram should indicate whether the reaction is exothermic or endothermic.

   c) Draw beakers of the two reactants before the neutralization and a beaker showing the products. Since water is a product of the reaction, you'll need to draw some water molecules in the final beaker (though, as usual, we will ignore the huge amount of water already present as the solvent).

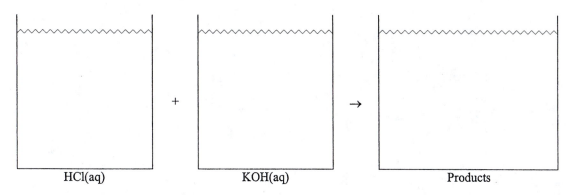

      HCl(aq)   +   KOH(aq)   →   Products

   d) The heat released during the reaction is primarily from the formation of what bond?

Continued on Back →

2) If it always takes energy to break bonds, then why are some reactions (for example, burning a candle) exothermic? Hint: If breaking bonds requires an input of energy, what happens when new bonds are formed?

3) Based on what you have learned about exothermic and endothermic chemical reactions, what is the major reason we get hot during strenuous physical activity?

Activity 9c: Determination of the Calories in a Peanut
COVER SHEET

Before Class
- Read the pages in your textbook dealing with "combustion," "calories," "Calories," and "heat capacity." Your instructor may assign specific pages for you to read.
- Complete the Pre-Activity Problem Set on the next page.

Key Concepts
- The specific heat capacity of a substance is the quantity of heat required to change the temperature of 1 gram of the substance by 1°C.
- Substances with large values of specific heat capacity resist changes in temperature.
- Water has an unusually high specific heat because of its relatively strong intermolecular forces.
- By observing the temperature change in a sample of water, we can determine how much heat was given off or absorbed by a chemical of physical change occurring in contact with the sample of water.
- 1 nutritional Calorie is equal to 1000 calories. (Note the use of the capital "*C*" when discussing *nutritional Calories*.)

Learning Objectives
- You will be able to utilize the magnitude of the specific heat capacity of a material to make predictions concerning the behavior of the material in response to the application or removal of heat.
- You will be able to calculate the amount of heat absorbed or released by a sample of water given the mass and temperature change of the water.
- You will be able to utilize the high specific heat capacity of water to explain climate patterns.

Pre-Activity Problem Set: Activity 9c

1) The following questions refer to placing an aluminum pot filled with 1500 g of water on the stove in preparation for cooking pasta.
    a) In addition to the water, what other things end up absorbing heat from the stove?

    b) Given that the specific heat capacities of water and aluminum are 4.184 J/g•C° and 0.90 J/g•C°, respectively, will the pot or the water show a greater change in temperature after 1 minute on the stove? Does your answer agree with your personal experiences? Explain your answers.

    c) How much heat (in joules) is needed to bring the 1500 g of water from room temperature (25°C) to its boiling point (100°C)?

2) Your friend moves from San Francisco to Washington, D.C. She writes in a letter to you about how the summers are much hotter and the winters much colder than back in California. You write back, relating what she is observing in the weather to what you are learning in your chemistry class about the specific heat capacity of water. What do you tell her?

CAUTION: Be sure to inform your instructor if you have peanut allergies. Follow your instructor's directions for lighting a bunsen burner. Be careful when working around the open flame.

Instructions:

1) If the fuel and water containers for the <u>soda can calorimeters</u> are not already constructed, follow the directions in Step #2 to make your own. If the fuel and water containers are already prepared, set up the calorimeter according to Step #3 (and the photo to the right).

2) To construct the fuel and water containers, you will need <u>two empty aluminum soda cans</u>. Using <u>scissors</u>, completely cut one of the aluminum cans 1 inch from the bottom. This will serve as the fuel container. Without cutting through the bottom of the fuel container, make a small indentation in the center of the fuel container to serve as a well to hold the peanut. Cut a second aluminum can 1 inch from the top. This will serve as the water container. Using a <u>hole-punch</u>, make two holes directly across from each other and 1 inch from the top of the water container. Dry the inside of the fuel and water containers.

3) To set up the calorimeter, obtain a <u>ring stand</u> and attach <u>two iron rings</u>. Insert a <u>glass rod</u> through the holes in the water container. Position the water container in the top ring, using the glass rod to suspend it.

4) To measure the Calories in a peanut, obtain <u>two shelled peanut halves</u> and place them in the fuel container. Record the weight of the fuel container and the peanut in the data table on the page 347. Using the graduated cylinder, add 100.0 mL of water to the water container of the calorimeter.

Stir the water with the <u>thermometer</u> and measure the initial temperature. Light the <u>bunsen burner</u> and, using the <u>tongs</u>, place the peanut halves in the flame. Once the peanut ignites, keep it in the flame for an additional 4 – 5 seconds to make sure it is fully burning. Transfer the peanut into the fuel container and raise the level of the iron ring until the flame touches the water container. This may take some patience in order to keep the peanut lit. Using the thermometer, gently stir the water. As soon as the peanut has stopped burning, measure the temperature of the water. After the fuel container has fully cooled (2 – 3 minutes), record the mass of the peanut residue and the fuel container. Dispose of the peanut and clean/dry the entire apparatus and repeat for a second trial.

	Trial #1	Trial #2
Mass of the fuel container and peanut before burning (grams)		
Volume of water (mL)		
Initial temperature of the water (°C)		
Final temperature of the water (°C)		
Mass of the fuel container and peanut after burning (grams)		

Questions (Show All Your Work):

1) How much peanut was burned (in grams) in each trial?

Trial #1:	Trial #2:

2) What was the temperature change of the water (in °C) in each trial?

Trial #1:	Trial #2:

3) What mass of the water (in grams) was used in each trial? Note: 1.00 mL water = 1.00 gram

Trial #1:	Trial #2:

4) How much heat was absorbed by the water (in J) in each trial?

Trial #1:	Trial #2:

5) How much heat was given off by the peanut (in J) in each trial?

Trial #1:	Trial #2:

6) Convert the heat given off by the peanut from J to Cal. Note: 1 Calorie = 4184 joules

Trial #1:	Trial #2:

7) Calculate the heat given off by the peanut (in Cal/gram of peanut burned) for each trial.

Trial #1:	Trial #2:

Name: _____ **Score:** _____ **points**

1) Attach your completed "Data and Calculations" page (*Part B*).

2) The nutritional label on a jar of peanuts indicates that the actual energy content of a 1.00-gram portion of peanuts is roughly 5.71 Calories. How does this value compare with your average experimental value for the two trials? Percent error can be calculated as follows:

$$\% \text{ error} = \frac{(\text{average experimental value} - \text{actual value})}{\text{actual value}} \times 100\%$$

3) There are many possible sources of error in this experiment (see sample answer below). List two additional sources of error that you can think of. What effect would each error have on the experimental value (too high, too low, or unclear)? Fully explain your answer.

> **Sample Answer**
> **Possible Error:** We accidentally wrote down a number that was too small when we initially weighed the unburned peanut.
> **Result of Error:** We will think that we burned more peanut than we actually did. When we divide the Calories released by the peanut by this large mass of burned peanut, we will get a final experimental value for the gross yield of the peanut that is too low.

Possible Error #1:

Result of Error #1:

Possible Error #2:

Result of Error #2:

Continued on Back →

4) Suppose you had two peanuts of equal mass and completely burned each in a separate calorimeter. The only difference between the calorimeters is that calorimeter "A" has 25 mL of water and calorimeter "B" had 150 mL of water.

a) Which calorimeter (A or B) will exhibit the greatest change in temperature? Explain.

b) Will changing the volume of water affect the results of the experiment in terms of the Cal/g of peanut burned? Explain.

5) The burning of 15.0 grams of vegetable oil causes the temperature of 250.0 mL of water in a calorimeter to rise from 22.5°C to 33.0°C. Determine the amount of heat (in Cal/g) given off during this combustion process. Show all your work.

Activity 9d: Kinetics and the Rates of Reactions
COVER SHEET

Before Class
- Read the pages in your textbook dealing with "kinetics" and "rates of reactions." Your instructor may assign specific pages for you to read.
- Complete the Pre-Activity Problem Set on the next page.

Key Concepts
- The speed at which a chemical reaction occurs is affected by the concentration of reactants, the surface area of any solid reactants, the temperature, and the presence of a catalyst.

Learning Objectives
- You will be able to discuss the factors that affect the rates of a reaction.
- You will be able to apply the factors that affect the rates of a reaction to various everyday scenarios.

Name: _____   Score: _____   points

1) Give an example of a chemical change that occurs very rapidly. Give an example of a chemical reaction that occurs very slowly.

   Fast Reaction: Slow Reaction:

2) Why does it make sense that increasing the concentration of reactant molecules can increase the rate of a reaction?

3) How does temperature affect the number of collisions between reactant molecules? Explain.

4) What is meant by the term *activation energy*? How can rolling a ball up and over a hill be related to activation energy?

5) What role does temperature play in helping reactants overcome their activation energy?

6) What is a catalyst? What role might a catalyst play in our automobiles? In our bodies?

Part A: Effect of Changing the Concentration of Reactants

CAUTION: 6M HCl is caustic. If any is spilled on your skin, immediately rinse with running water and inform your teacher. Wear your safety goggles at all times during this activity.

Instructions: Place a small <u>piece of chalk</u> in each of the <u>two watch glasses</u>. Add 10 drops of <u>0.5M HCl(aq)</u> to one sample of chalk and <u>6M HCl(aq)</u> to the other. Record your observations. Dispose of waste according to instructor's directions.

Observations:

Questions:
1) Explain the differences in your observations in terms of the concentration of H^+ ions present in the two different acid solutions.

2) Marble, like chalk, is composed of $CaCO_3$. Explain why monitoring the acidity of rainfall would be important with regard to conserving important outdoor buildings and monuments.

Part B: Effect of Changing the Surface Area

CAUTION: Clear flammable materials from lab bench. Never direct the burst of coffee creamer at another person or at any flammable material. Wear your safety goggles at all times during this activity.

Instructions: After lighting the <u>candle</u>, use a <u>spatula</u> to hold a small amount of <u>coffee creamer</u> in the flame for 5 seconds. Record your observations. Next, draw a small amount of coffee creamer into a <u>disposable pipette</u>. While standing at arm's length from the candle, aim a burst of coffee creamer at the flame. Record your observations. Wipe down your lab bench when you are done.

Observations:

Question:
1) Explain your observations in terms of surface area of coffee creamer exposed to oxygen and the flame.

Part C: Effect of Changing the Temperature

CAUTION: Be careful when using hot plates. Use tongs or heat-resistant gloves to handle hot glassware. Wear your safety goggles at all times during this activity.

Instructions: Add 25 mL of 0.2 M $CuSO_4$ to each of two 50-mL beakers. Cool one of the beakers in an ice bath. Heat the other beaker on a hot plate until it reaches 80°C on a thermometer. Carefully remove the beaker from the hot plate using tongs or heat-resistant gloves. Use a spatula to add a few pieces of granular zinc to the hot solution. Record how long it takes for any color changes to occur. Repeat by adding a few pieces of zinc to the cold solution. Again, record your results. Dispose of waste according to your instructor's directions.

Observations:

	Hot Solution	Cold Solution
Time for 1st color change in the...		
Time for 2nd color change in the...		

Question:
1) What effect does increasing the temperature have on the rate of reaction? Your answer should address the effect of heat on submicroscopic particles in terms of the frequency and the force of their collisions.

Part D: Modeling the Significance of the Orientation of Collisions

Instructions: Place two of the Styrofoam balls with 6 pieces of Velcro on them in a box or deep tray. Gently shake the container until the two balls stick. Repeat with two balls that have only 1 Velcro square. Describe how readily the balls stick together in each case.

Observations:

Questions:
1) In which case (the balls with 1 Velcro or 6 Velcro pieces) was the required orientation for successful collision (forming a product "molecule" from the two individual "atoms") more restrictive? Explain.

2) Did this agree with your observations of how long it took for the balls to stick? Explain.

Part E: Effect of Adding a Catalyst

Instructions: Using <u>two 50-mL beakers</u>, warm two, 10-mL samples of <u>0.5% corn starch solution</u> to 40°C. In a <u>test tube</u>, collect 2 mL of <u>saliva</u> from yourself and your lab partner. Using a <u>glass rod</u>, thoroughly mix the saliva into one of the corn starch/iodine solutions. The other beaker, without saliva, will serve as the control for comparison. Wait 5 minutes and add 10 drops of <u>iodine solution</u> to each beaker. Record your observations. Dispose of the waste according to your instructor's directions.

Observations:

Questions:

1) Iodine turns dark blue in the presence of starch. Which beaker (the one with or the one without saliva) has more starch present? Explain.

2) What role did the saliva play in breaking down the starch? Briefly explain.

Name: _____   **Score:** _____ **points**

1) As a reaction progresses and the reactants are consumed, will this tend to increase or decrease the rate of reaction? Explain.

2) Powdered sugar dissolves in water faster than the same mass of large sugar crystals. Explain why.

3) As scientists, we often want to speed reactions up. However, there are certain processes that we might want to slow down, for example, how fast iron rusts, how quickly we age, or the speed at which food spoils. At home we can slow down the rate at which food goes bad by tightly sealing the food in plastic wrap and placing it in the refrigerator.
   a) In terms of the factors that affect the rate of a reaction, what role does the plastic wrap serve? Explain.

   b) What role does the refrigerator serve? Explain.

4) Using as a model trying to hit a home run in baseball, explain the factors that determine whether a successful chemical reaction will occur.

Activity 9e: An Introduction to Electrochemistry
COVER SHEET

Before Class
- Read the pages in your textbook dealing with "oxidation and reduction." Your instructor may assign specific pages for you to read.
- Complete the Pre-Activity Problem Set on the next page.

Key Concepts
- During some chemical reactions, one or more electrons are transferred from one reactant to another. Such reactions are called oxidation – reduction reactions.
- The electricity produced from a battery comes from chemical reactions.
- Electrolysis uses electrical energy to produce a chemical change.

Learning Objectives
- You will be able to define basic terms used in electrochemistry.
- You will be able to use experimental results to rank metals according to how reactive they are.
- You will be able to draw a voltaic cell.
- You will be able to build a working voltaic cell.
- You will be able to explain, in simple terms, how a battery works.

Name: _____   Score: _____ points

1) Define the following terms:
   a) Oxidation

   b) Reduction

   c) Cathode

   d) Anode

2) The questions below and on the following page refer to the oxidation – reduction reaction between solid magnesium metal and aqueous lead(II) ions.
   a) Write the balanced half-reaction for the oxidation of solid magnesium metal to aqueous magnesium ions. Hint: Use enough electrons in the balanced half-reaction so that the total charges on each side of the reaction are the same.

   b) Write the balanced half-reaction for the reduction of aqueous lead(II) ions to solid lead metal.

   c) Combine the reactions from Questions #2a and #2b to give the balanced reaction for the formation of magnesium fluoride from its elements.

Continued on Back →

d) Use the fact that magnesium is a more active metal than lead to complete the following drawing of a voltaic cell made out of magnesium and lead. Be sure to label the following parts of your drawing: the salt bridge, the lightbulb, the anode, the metal at the anode, the reaction at the anode, the cathode, the metal at the cathode, the reaction at the cathode, the oxidation reaction, the reduction reaction, and the direction of the electron flow.

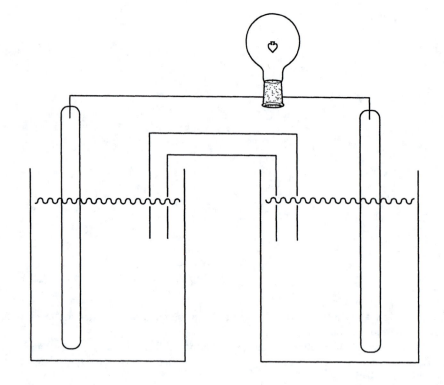

Instructions: If the zinc and copper wires are not already prepared, begin by cleaning them with steel wool and cutting off about 5 cm using wire cutters. The wires should not be insulated. Fill two wells of a well-plate—one with $1.0 M Cu(NO_3)_2(aq)$ and a second with $1.0 M Zn(NO_3)_2(aq)$. Place the zinc wire in the well with the copper solution and the copper wire in the well with the zinc solution. Record your observations below. If nothing happens, write "no reaction." Rinse, dry, and save the metal wires for *Part B*. Dispose of the solutions according to your instructor's directions.

Observations:

Zinc metal in solution of copper ions:

Copper metal in solution of zinc ions:

Questions:

1) For the reaction that you saw occur:
   a) Write the half-reaction for the oxidation of the metal.

   b) Write the half-reaction for the reduction of the metal ions.

   c) Using your answers to Questions #1a and #1b, write the total reaction that occurred in the well-plate between the metal and the metal ions.

   d) The energy that accompanied the reaction in Question #1c was lost. Explain how you could have set up the reaction to harness the electrochemical energy to do work.

2) For the reaction that did not occur:
   a) Write the half-reaction for the oxidation of the metal.

   b) Since the reaction in Question #1a occurred and the reaction in Question #2a did not occur, what can you conclude about which metal is more active (and is better at giving up its electrons)?

Instructions: Clean the zinc and copper wires from *Part A* with steel wool. The wires will serve as your electrodes. Half fill one 50-ml beaker with 1.0 M $Cu(NO_3)_2(aq)$ and half-fill a second 50-mL beaker with 1.0 M $Zn(NO_3)_2(aq)$. Use scissors to cut a 2-cm-wide strip of filter paper to serve as your salt bridge. Saturate the strip of filter paper using a dropper bottle with 1.0 M $NaNO_3(aq)$ and use it to bridge the two beakers. Make sure the salt bridge touches the solution in both beakers and that it does not dry out during the experiment. Place the zinc wire so it hangs into the beaker with the solution of zinc ions and the copper wire so it hangs into the beaker with the solution of copper ions. Use a multimeter to measure the voltage across the zinc and copper electrodes. After you have finished, dispose of the cell contents according to your instructor's directions.

Questions:
1) Based on the results from *Part A*,
    a) Which metal (Zn or Cu) is being oxidized in your cell? Explain.

    b) Which ions (Zn^{2+} or Cu^{2+}) are being reduced in your cell? Explain.

2) Use your answers to Questions #1a and #1b to draw a voltaic cell made out of zinc and copper (refer to the model from Question #2d in the Pre-Activity Problem Set). Be sure to label the following parts of your drawing: the salt bridge, the lightbulb, the anode, the metal at the anode, the reaction at the anode, the cathode, the metal at the cathode, the reaction at the cathode, the oxidation reaction, the reduction reaction, and the direction of the electron flow.

Part C: The Lemon Battery

Instructions: Prepare <u>two lemons</u> by pressing down on them and rolling them back and forth on the desk. Press hard enough to squeeze the lemon, but not hard enough to break the peel. If the wires are not already prepared, begin by using <u>wire cutters</u> to cut off <u>three 8-inch lengths of insulated copper wire</u>. Use the wire cutters to strip off about 2 inches of insulation from both ends of all three wires. Wrap the end of the first length of copper wire around a <u>galvanized nail</u> and the other end around <u>a penny</u>. Press the nail into one <u>lemon</u> and the edge of the penny about half way into a <u>second lemon</u>. Attach one end of the second wire to a <u>second nail</u> and insert the nail into the lemon that already has the penny. The nail and the penny should be about 2 cm apart and should not touch. Attach one end of the third wire to a <u>second penny</u>. Insert the penny into the lemon that already has the nail stuck in it. Test your battery by connecting the two free wires coming from the lemon battery to an LED or the contact on the battery holder of a small digital clock or calculator. If the test object does not work, switch the two wires, since they may be attached to the object in reverse. Additional lemons can be added to the sequence to increase the voltage. Carefully look at the lemon where the nail and penny are inserted. Record your observations.

Observations:

Questions:
1) Unlike the cell you built in *Part B*, there are no copper ions to be reduced, rather it is the acid in the lemon acting as a source of H^+ ions that can be reduced. Answer the following questions knowing that galvanized nails are coated with zinc and that pennies contain copper.
   a) What other fruits or vegetables do you think would make a good battery? Briefly explain.

   b) Using your observations as a clue about what the product might be, write the balanced half-reaction for the reduction of the H^+ ions.

   c) Knowing from *Part A* whether zinc (from the nail) or copper (from the penny) is the more active metal, write the balanced half-reaction for the reduction of the metal that is occurring in the lemon battery.

Instructions: Half-fill a small beaker with water. Add a few drops of phenolphthalein indicator and a scoop of washing soda (sodium carbonate) and stir until most of the salt dissolves. If the wires are not already prepared, begin by using wire cutters to cut off two 8-inch lengths of insulated copper wire. Use the wire cutters to strip off about 2 inches of insulation from both ends of both wires. Take four alligator clips and attach one to each end of both wires. Use the alligator clips to attach the wires to two pencils that have been each been sharpened at both ends. The graphite in the pencils will serve as our electrodes. Place the free ends of the pencils in the beaker of water. Look carefully at the electrodes in the beaker so you can tell if there is a difference later. Using the loose alligator clips, attach the pencil electrodes to the terminals of the 9-V battery. When you are done recording your observations, disconnect the electrode from the battery.

Observations:

Questions:

1) We have all used batteries to run flashlights, toys, or radios. Here, the energy from the battery is being used to take apart water molecules. Write the balanced reaction, including states, for the decomposition of water molecules into their two constituent elements.

2) If you were to collect the gases being generated at each electrode, you would find that the cathode is generating twice as much gas as the anode. Which gas is being generated at the anode? Which is being generated at the cathode? Explain your answers based on the balanced equation from Question #1.

    Anode Gas: Cathode Gas:

3) Draw a potential-energy diagram for the decomposition of water (Question #1). Label the reactants and products. Your diagram should clearly indicate whether the reaction is endothermic or exothermic.

Post-Activity Problem Set: Activity 9e

Name: _____   Score: _____   points

1) Using the Internet or your textbook, answer the following questions referring to hydrogen – oxygen fuel cells. You will need to record the source of your information.

   a) What is the reaction occurring at the anode in a hydrogen – oxygen fuel cell?

   b) What is the reaction occurring at the cathode in a hydrogen – oxygen fuel cell?

   c) What is the total oxidation – reduction reaction occurring in a hydrogen fuel cell?

   d) What is so attractive about the products formed by a hydrogen fuel cell?

   e) Where are hydrogen fuel cells currently being used?

   f) Source of information:

2) The following questions refer to the electrolysis of water that you performed in *Part D*.

   a) There are two products formed at the anode. One is the gas you identified in *Part D*, Question #2, and the other is the H^+ ion. Write the balanced half-reaction, including electrons, for the reaction occurring at the anode.

   b) There are also two products formed at the cathode. One is the gas you identified in *Part D*, Question #2, and the other is indicated by the pink color of the phenolphthalein. Write the balanced half-reaction, including electrons, for the reaction occurring at the cathode.

Continued on Back →

3) Use the results of the following experiments to answer the questions below.

Experiment	What Was Tested	Did the Reaction Occur?
#1	Sn(s) was put in Ag^+(aq)	Yes
#2	Ag(s) was put in Sn^{2+}(aq)	No
#3	Fe(s) was put in Sn^{2+}(aq)	Yes
#4	Sn(s) was put in Fe^{2+}(aq)	No

a) What is the oxidation – reduction reaction that occurred as a result of Experiment #1?

b) What is the oxidation – reduction reaction that occurred as a result of Experiment #3?

c) Rank the metals from most active to least active. Remember that the most active metals are the ones that readily give up their electrons.

Ranking: (Least Active) _____ < _____ < _____ (Most Active)

d) Would you expect to observe a reaction if Fe(s) was placed in Ag^+(aq)? Explain.

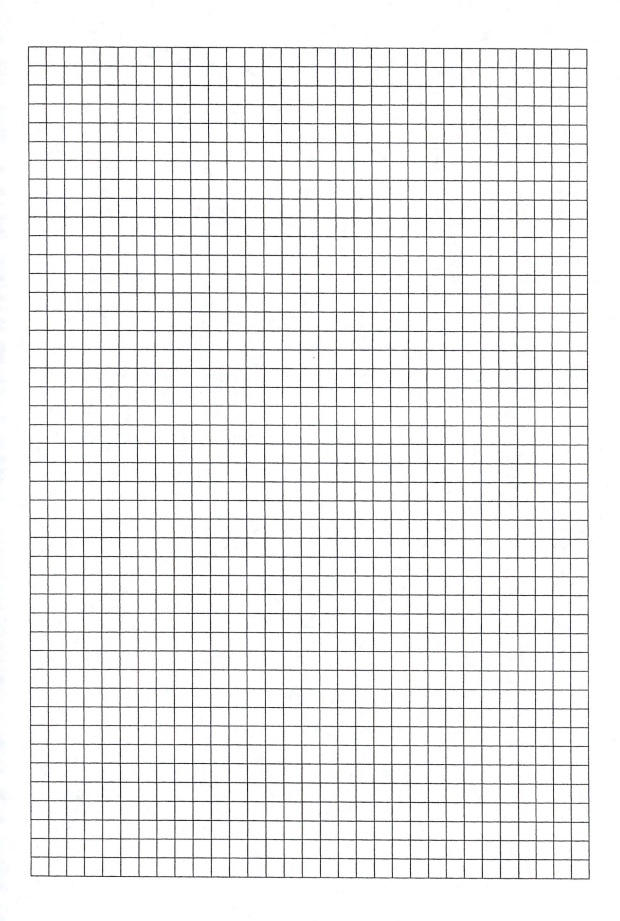

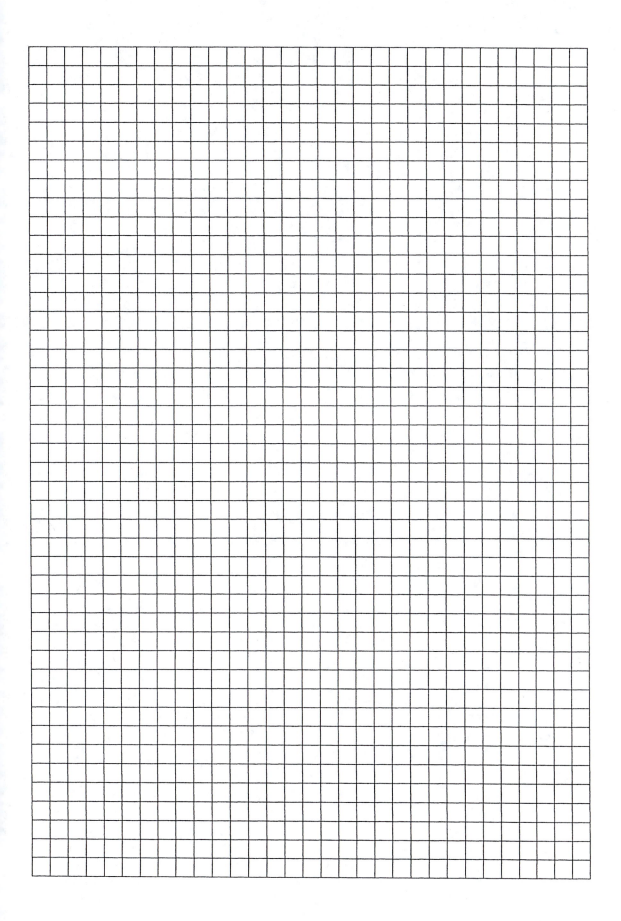

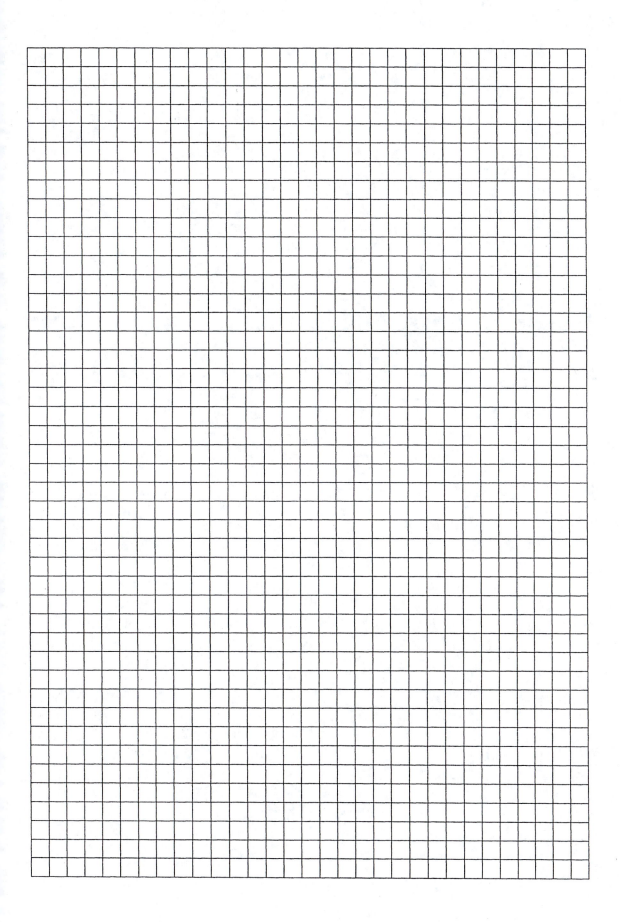

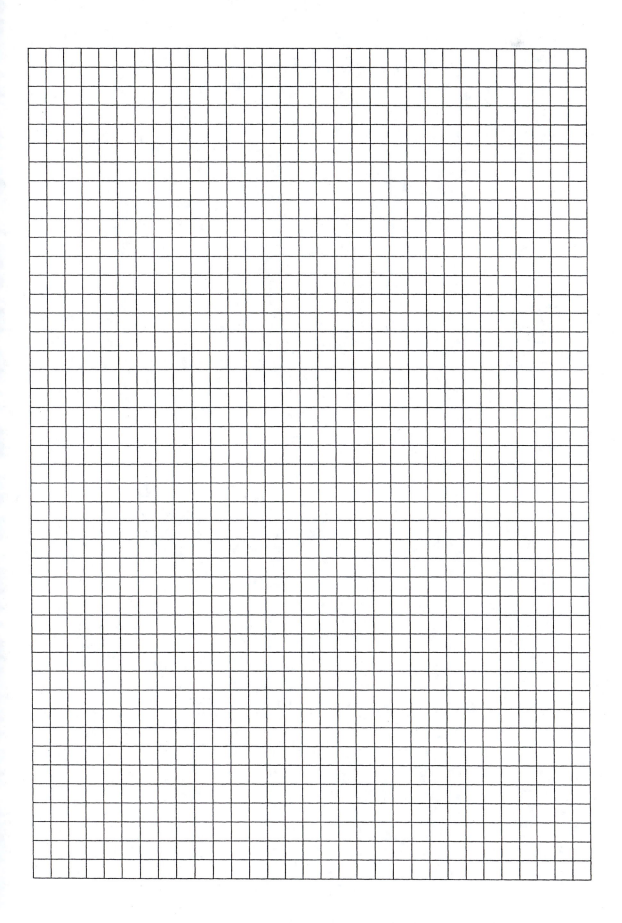

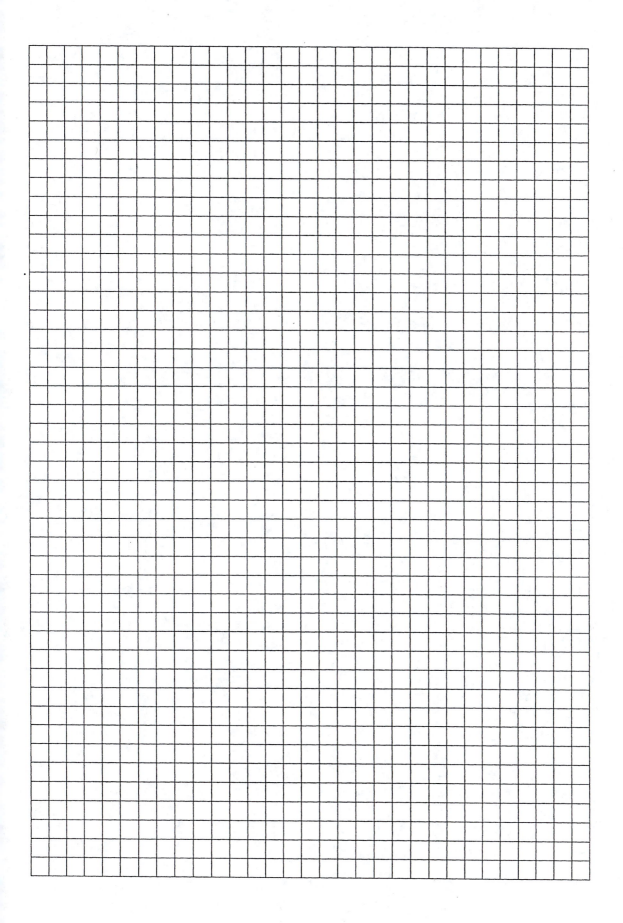

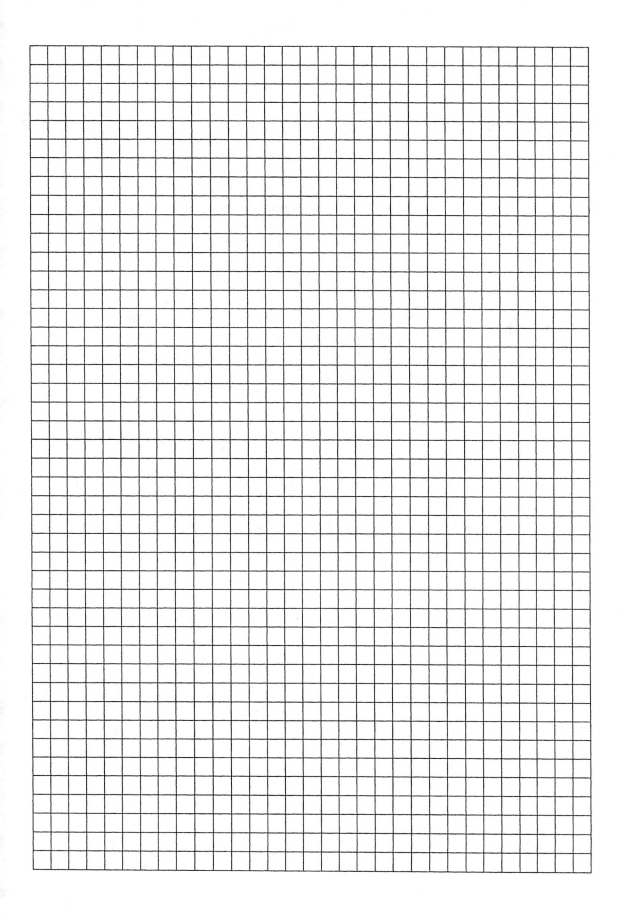